CANON
PRODUCTION SYSTEM

CANON
PRODUCTION SYSTEM

Creative Involvement of
The Total Workforce

Compiled by the Japan Management Association

Translated by Alan T. Campbell

Edited by Constance E. Dyer

Preface by Norman Bodek, President, Productivity, Inc.

Productivity Press
Cambridge, Massachusetts   Norwalk, Connecticut

Productivity Press　　　　　　　Productivity, Inc.
P.O. Box 3007　　　　or　　　　Merritt 7 Corporate Park
Cambridge, MA 02140　　　　　　101 Merritt 7, 5th Floor
(617) 497-5146　　　　　　　　Norwalk, CT 06851
　　　　　　　　　　　　　　　　(203) 846-3777

Library of Congress Catalog Card Number: 85-63498
ISBN: 0-915299-06-2

Cover design: Russell Funkhouser
Typeset by Rudra Press, Cambridge, Massachusetts
Printed and bound by Arcata/Halliday
Printed in the United States of America

Library of Congress Cataloging-in-Publication Data

Kyanon no seisan kakushin. English.
　　Canon production system.

　　Translation of: Kyanon no seisan kakushin.
　　Includes index.
　　1. Kyanon Kamera Kabushiki kaisha — Management — Employee participation.　2. Camera industry — Japan — Management — Employee participation — Case studies.　3. Production management — Japan — Case Studies.
I. Dyer, Constance E.　II. Nihon Nōritsu Kyōkai.
III. Title.
HD9708.5.C354K92513　1987　　681'.418'0683　　85-63498
ISBN 0-915299-06-2

90 10 9 8 7 6 5

Publisher's Preface

In the past few years, I've been to Japan 15 times on study missions with top American executives to find out what Japanese companies are doing to become world leaders in quality and productivity.

For a long time we didn't understand the incredible transition taking place in these small islands off the coast of China. Too many of us held on to the belief that goods made in Japan were inferior. Perhaps our victory in the war gave us a false sense of superiority that led us to overlook the creativity and ingenuity of the Japanese people.

The Japanese success story is an astounding one — and much of it is reflected in this book. What are the Japanese doing? We can see Canon's technological advances, for example, in high-precision cameras, optical equipment, laser technology, and business machines. To attribute their success to superior product development alone, however, would be naive. The United States has also proven itself capable of technological innovation and development — putting a man on the moon, for example, developing the transistor, and creating much of the manufacturing technology relied upon by Japanese companies today.

So there must be more to the magic evolving in Japan. What is it we've missed? Technology, product development, and diversification are part of the picture, but more important is a new management style and what we can call the "new worker," more responsible and fully integrated in new, highly efficient production systems. It is an approach that is subtle, yet incredibly powerful in its ability to promote rapid change and growth.

Canon, as much as any other company, illustrates the tremendous potential for achievement represented by this new approach. Fifteen years ago Canon was struggling to produce a camera that would appeal to the world market. The finest cameras were made almost exclusively in Germany, and Kodak had given us photography at a price everyone could afford — the little Brownie Kodak camera. Canon revolutionized the industry, however, giving us the latest in technological advances and precision at affordable prices.

This book is not about Canon's products, however, but rather the streamlined production and management system that inspire Canon people to deliver high quality as well as the high productivity levels that keep prices down. Ian Brammer, managing director of Technology Transfer Council of Australia, describes Canon's strategy aptly, when he calls the Japanese secret "continuous improvement through the elimination of non-value-added wastes by the creative involvement of all employees."

Waste elimination, continuous improvement, and a vitalized workforce — these are Canon's keys to success. Its first step was the elimination of all non-valued-added wastes: overproduction, excess inventory, ineffective planning and operations, and wasted human resources — the creativity and ingenuity of the line worker.

This last element is important, because the tradition of lifetime employment in Japan has forced major Japanese corporations like Canon to improve quality and productivity without letting people go. As a result, workers are used more creatively and given more responsibility, and Japanese companies in general have tended to diversify — to support a permanent workforce by creating more wealth.

How can we manage people, technologies, and production facilities to diversify and deliver higher and higher value-added products — without increasing the cost? The story of how Canon is accomplishing this goal should be of great interest to American manufacturers. Canon confronted problems common to both American and Japanese manufacturers and came up with simple solutions to improve production facilities and manage a total improvement process. Using the principles of just-in-time developed by Toyota, Canon designed efficient production systems to suit their own product lines and manufacturing conditions. They learned how to set demanding company goals and actually achieve them, including the goal of increasing productivity by 3 percent per month throughout the organization — which they achieve without laying off a single permanent worker!

Like other top Japanese companies, Canon has invested a great deal to develop the creative talents of their workforce. At the Canon Toride plant where they manufacture copiers, I saw the most motivated workforce I've ever seen in my life — men and women fully integrated with sophisticated robotics in a smooth continuous flow production line. Last year, Canon received 50 ideas per worker to improve their job and the company — just imagine getting one new improvement idea every week from every employee. This is an incredible statistic when you consider that the average American company may get one creative idea per worker per year or once every 5 years.

I'm pleased as a publisher to bring this book to America. I want to acknowledge the contribution made by the Japan Management Association whose authors compiled the original Japanese text from material received directly from Canon. I also want to thank the people who helped create this English version — the translator, Alan Campbell; Connie Dyer, who edited and expanded the text and figures; Patty Slote, who coordinated the book's production; Russ Funkhouser, who designed the cover; Marie Kascus, who provided the index; and the staff of Rudra Press — Nanette Redmond, Ruth Knight, Caroline Kutil, Laura Santi, and Leslie Goldstein. I especially want to thank Ryuzaburo Kaku, President of Canon, for the cooperation of all the people at Canon who helped us with this project, with special thanks to Masayoshi Hiramatsu of Canon's Public Information Office, who gave us many hours of his time to answer our questions and bring us up to date on the Canon Production System.

Norman Bodek

Editor's Preface

Canon's continuous improvement system is considered one of the best in Japan. Visit any Canon factory and you'll quickly see why. In every part of the plant there are visible signs of ongoing improvement activity: up-to-date status charts in every work center, and eye-catching handmade posters promoting the factory-wide improvement focus for the month. A small group meeting is held in the lunch room, while in another room workers visiting from another plant view improvement results presented on an overhead projector. In assembly, a young foreman consults a long list and visits with workers — he's distributing small cash awards for work improvement proposals to almost every worker on the line! Rigorous control activities are also in evidence. Clear work guidelines and quality standards are posted at every work station; quality and productivity measures are recorded continuously. And throughout the plant, a quiet intensity prevails — everyone works at a rapid pace, but there's no sign of a rush anywhere. This is the result of Canon's "software for success" — the "hidden" management systems that make the idea of continuous improvement a functioning reality at Canon.

In 1975, Canon set an ambitious goal for itself — to become a world-class corporation within six years. Looking back now, its success might appear predictable — diversification and development within areas of technological strength, combined with astute, well-timed marketing decisions. In this book, however, the keys to Canon's success are explained from the perspective of those who contributed inside the factories — the managers, engineers, and workers whose small improvements every day made large scale improvement possible. This perspective is valuable to us. Read between the lines and you can see that Canon had to overcome the same problems American manufacturers face at the line and middle-management levels: ineffective communication, over-specialization and isolation, lack of trust at operational levels, and the high costs of lack of direction and constant change at the design and trial production stages.

Any company planning to make major changes in operations must confront these and other internal obstacles — established patterns of thinking and behaving that strangle the very best ideas and intentions. As one American manager put it, "We know what changes to make and why; and we know it can work — but how do you get several thousand people involved and cooperating to make it happen?" Japanese managers ask the same question — that's why The Canon Production System was published in Japan.

Canon is an innovator in many ways. It was the first camera manufacturer to use conveyors in camera assembly, and it followed Toyota's lead in implementing just-in-time production. As Canon learned in the mid-70s, however, new technologies and strategies require management systems ("software") that not only promote rapid change but also maintain improvement levels during that change.

Canon recognized that the rapid improvement it sought could not be accomplished without common company goals and continuous, effective communication. To facilitate comunication and cooperation, the company was reorganized as a matrix of product groups and management systems for development, production, and marketing. Study groups formed at every inter-section of the matrix and brought together previously isolated managers, technologists, and direct and indirect personnel to design and implement the new systems. Basic strategies for im-provement were introduced — drives for waste elimination, standardization, and reliable planning and control — to give each group a common language and a common approach to fundamental improvement.

Chief among these strategies was effective planning. At Canon, making a good plan — i.e., setting realistic, concrete, and detailed objectives — is only the beginning; making it hap-pen is the test. And it is here that Canon's management software reveals its power. "Plan = Results" is the objective, and Canon has achieved a surprising balance between planning and doing.

Effective planning and follow-through can't happen with-out teamwork and involvement, however. What brings Canon workers into this intense process; what keeps them involved? This is where the creativity and ingenuity of Canon managers have really made a difference. The terms "promotion" and "im-provement management" are used throughout the book, and they refer to that aspect of planning that has to do with "making sure it happens" — through the creative involvement of every worker. A manager must plan how to achieve production and improvement targets. He must also plan promotional activities that involve everyone actively in achieving those targets. Small group activities, the work improvement proposal system, an-nouncements and awards, the various improvement drives and campaigns — these are the formal strategies for involvement. It is creative management and interpersonal skills of individual foremen and factory managers, however, that sustain the vitality of individual Canon work centers and contribute to the continu-ous high rate of productivity improvement the company has enjoyed for the last ten years.

This book offers a unique view inside Japanese manufac-turing — one that won't be found in other books on just-in-time production or Total Quality Control. It is a casebook of simple day-to-day techniques Canon managers used to apply those broader principles successfuly in their workplaces. The Japanese are masters of the art of kaizen or continuous improvement, and this book more than any other captures the spirit of that never-ending commitment to quality.

Constance E. Dyer

Contents

Introduction to the
Japanese Edition

Management — rather than national or cultural environment — is the principal factor in ensuring efficient, high-quality production.

Ryuzaburo Kaku, President
Canon, Inc.

Canon excels in product development. In 1961, when mid-price cameras were selling for around $63.00, Canon achieved an unprecedented price of under $52.00 with the Canonet. Similar examples followed — the AE-1, an electronic single-lens reflex camera that dominated its day, and the PC-10 copier, introduced for personal use in 1982. The excitement created in the market by the specifications and functions of these products is well documented.

Canon's strength in product development provides one explanation for the company's high growth rate. Another factor has been drastic reduction in manufacturing costs achieved through automation and the use of molded parts, labor-saving measures, and other innovative production techniques.

Notwithstanding these factors, a chance failure in product development in 1975 led Canon to completely restructure its corporate management systems. In a new matrix organizational framework, three product groups were linked by three management systems to promote improvement in product development, production, and marketing.

This book takes a close look at one of those drives for improvement — the Canon Production System (CPS). From 1976 to the present, CPS has earned Canon just under $25 million per year in profits resulting from the elimination of waste in all areas of manufacturing: work-in-process, defects, physical facilities, expenses, indirect labor, design, talent, motion, and startup (mass trials). CPS has introduced successful methods of goal achievement planning in all areas of production, and continues to evolve as a management tool.

The more carefully manufacturing operations are planned and executed, the lower product cost and the higher product quality can be. Thus, CPS systems operate as a form of total quality control. Productivity improvement is a particularly important part of Canon's manufacturing operations; it is promoted by CPS management "software," working together with Canon's "hardware" manufacturing technologies.

The CPS focus on manufacturing efficiency might also be viewed as a quality-of-work-life (QWL) program, because it has given factory employees more control over their own work. CPS coordinates the ZD (zero defects quality control) program, the work improvement proposal system, small group activities, and many other employee-based improvement activities.

This book is about Canon's approach to managing improvement; but it is also a partial record of practical achievements for the period 1976-1982, drawing on the experiences of different Canon factories. Change and improvement at Canon did not occur as uniform, across-the-board implementation of techniques developed in a central office. Rather, CPS fostered independent improvement activities in every factory, and then served as a clearinghouse to promote effective sharing and transfer of knowledge throughout the system. In this way, workers in every factory, in each line section, have come to view improvement as their own unique contribution to Canon, as well as their own important responsibility within the system. The secret of the Canon Production System's success lies as much in those individual efforts as in the management system that ties them all together.

We dedicate this book to anyone actively engaged in manufacturing improvement efforts and hope it will prove useful.

We gratefully acknowledge the interest shown by Canon, Inc. and President Kaku during the preparation of this book. We are especially grateful to CPS Chairman and Managing Director Yamagata, as well as to individuals in the CPS promotion department and the factory CPS offices. Their active cooperation made this book possible.

The Compilers
Japan Management Association

CANON
PRODUCTION SYSTEM

1

The Birth of the Canon Production System

We must move away from the kind of hardware thinking that relies heavily on products toward something like software thinking, which emphasizes the structure of operations.

Ryuzaburo Kaku, President

In 1975, Canon encountered obstacles in the development and sale of calculators and went into the red for the first time since its stock was listed in 1949. This, together with the oil crisis, caused Canon to record a deficit of 200 million yen ($665,000) for the first half of the year. (Figure 1-1) Although the company recovered quickly, making up the deficit by a large margin during the second half, the temporary inability to pay a dividend was a great shock to company executives. At a closed-door meeting later in the year, a new approach was adopted — not merely to reorganize, but to build morale. The company was given a goal: to become a world-class corporation in one great leap.

President Ryuzaburo Kaku (then Managing Director) (Photo 1-1) proposed a six-year "Premier Company Plan," beginning in 1976. Under the plan, Canon had three years to become a leading corporation in Japan and another three to become a world leader. The new plan was introduced along with more immediate measures for recovery — reduced operations, cost curtailment and efforts to strengthen calculator sales and the newer camera and copier product lines. Goals were established: 1) to strengthen the company's financial position (an operating profit rate of 15 percent with no debt); 2) to raise levels of development, production, and marketing (in both quantity and quality); and 3) to achieve a reasonable growth rate (a 30 percent rate of increase in market share and a 15 percent rate of increase in sales).

Photo 1-1 Ryuzaburo Kaku

MOVING TOWARD SOFTWARE THINKING

In the past Canon often moved ahead of its competitors by introducing innovative, attractive products, only to end up in second or third place because of ineffective long-range planning. Management tended to be on a day-to-day basis, and the company was unable to take advantage of opportunities for growth. To overcome this inflexibility, the Premier Company Plan outlined a series of objectives to strengthen the corporate structure and increase its operational efficiency.

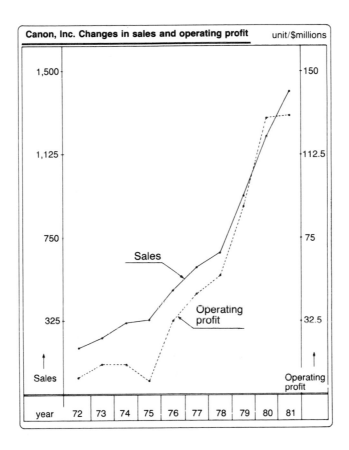

Figure 1-1 Changes in Sales and Operating Profit

The first task was to simplify Canon's complex production system of interlocking and interdependent factories. Before the new plan, a typical product was processed at plant A, transported to plant B for subassembly, and then taken to plant C for final assembly. Canon's goal was to bring these functions together in individual plants.

A second, related task was to create a new organizational framework based on the product groups — camera products, business machines, and optical products. This would strengthen management and develop human resources by concentrating areas of specialization that had developed as product lines diversified. Canon's goal was to build strength by creating areas of independent responsibility as well as expertise. For example, in the product groups system, marketing was made an independent company to increase its effectiveness and eliminate crosscurrents between manufacturing and sales.

Three management systems were established to help make these changes and coordinate common functions in all areas of the company: the Canon Production System (CPS), the Canon Development System (CDS), and the Canon Marketing System (CMS). To develop these systems, three study committees were organized with company-wide members. For example, members of the Canon Production System Study Committee included all Canon plant managers and division chiefs and the presidents of affiliated manufacturers. Its chairman was the managing director in charge of production.

The system study committees were asked to build management systems based on long-range planning that would give Canon the ability to move ahead quickly and to withstand any unforeseen market changes. The development system (CDS) committee was to foster the development of new products, high in performance and quality, matched to consumer needs and timed for the market. The production system (CPS) committee was to eliminate waste so the company could manufacture higher-quality products at lower costs and withstand both a stronger yen and intensified competition. The marketing system (CMS) committee was expected to expand and strengthen Canon's independent domestic and overseas sales network by building a high-quality service and sales force. Separate offices were established in company headquarters to promote the work of each committee.

MATRIX MANAGEMENT AT CANON

A matrix management structure evolved out of the interaction between the product groups and the systems committees. (Figure 1-2) Typically, a matrix organization combines departments organized by function with groups organized by project.

In Canon's case the committees are investigative bodies that are permanently provided with offices. The product groups form the vertical axes of the matrix; the Canon systems (along with central staff, i.e., personnel, accounting, etc.) represent the horizontal links between each group in R&D, production, and marketing. The relationship between the product groups and the Canon systems can be explained by a farming analogy. The product groups discover a strong and tasty variety of produce, then sow the seeds, and harvest the crop. The Canon systems improve the soil and the environment — they provide the fertilizer and irrigation; they prepare the soil and keep it free of weeds, and so forth.

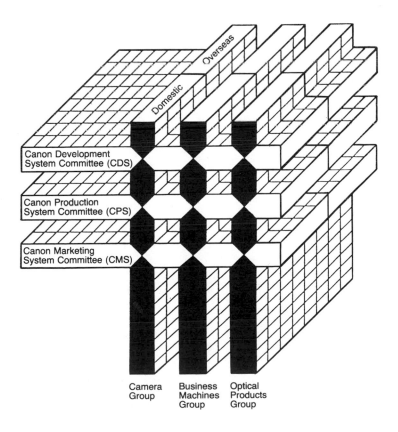

Figure 1-2 The Canon Matrix

The Canon style matrix — independent businesses and plants organized to work toward common objectives — is unique for a Japanese company. The key to its success is less control and more support by central offices and the efficient transfer of successful techniques to other plants. In this respect the role of the Canon Production System (CPS) is central — to help develop successful productivity and quality improvement strategies, management techniques, and workplace vitalization schemes and to coordinate their implementation throughout the company.

One unavoidable result of the matrix system is that actual operations on the line must be carried out under the direction of two "bosses," e.g., product and systems managers. This has not been a serious problem at Canon, however. One reason is that the system study committee chairmen are managing directors already deeply involved in staff operations. Another is that the product groups have benefited greatly from the study committees' activities, e.g., the HIT production system (the CPS version of just-in-time parts supply) and TS 1/2 (a CDS program to cut product development time in half).

CPS GOALS AND STRATEGIES

The CPS Study Committee established three company-wide goals: "To strive for the best quality, the lowest cost, and the fastest delivery anywhere." These goals were to be achieved pursuing three basic strategies that were already part of Canon's production philosophy.

Reliable Quality Assurance

Canon had always been known for the high quality of its products. The committee recognized, however, that to maintain public confidence in the company during the reorganization and in the future, top priority had to be given to developing a truly reliable quality assurance system.

The Ultimate Continuous Flow System

Canon was the first precision industry to change from lot assembly to the conveyor belt system. This increased efficiency and simplified operations. Moreover, since the status of operations could be seen at a glance, it became easier to control productivity, work-in-process, and defects. Canon's goal was to create a single continuous flow system that included all aspects of production — not just assembly and related processing operations, but also indirect operations such as paper documentation and reporting.

Effective Development of Human Resources

Most system problems turn on a human factor. And if problems are not pursued far enough to find that factor, solutions will be inadequate. Canon discovered that problems are solved more effectively when the people involved are given opportunities to develop and grow. Effective planning for human resources development can bring to the surface each individual's desire to work and improve and make it serve the company as well as the individual. All CPS activities are organized around this simple principle.

Emphasis on human resources development is a fundamental part of Canon's corporate philosophy that predates the inauguration of CPS. It can be found in the "Four Canon Principles" enunciated by Canon's founder, Takeshi Mitarai. These principles promote merit-based advancement, family spirit, health, and "the three selfs" — self-motivation, self-respect, and self-reliance. (Figure 1-3)

This philosophy has always guided company policy at Canon. For example, distinctions between line and staff workers were eliminated before World War II. Advancement is based on testing, and no discrimination by sex or educational background is permitted. The mandatory retirement age has been extended several times, with a current policy of reemployment for three years after age 60.

- "Three-Self" Policy
  We will proceed in the spirit of self-motivation, self-respect, and self-reliance.

- Merit System Policy
  We will maximize our human resources through the merit system.

- New Family Policy
  We will cooperate to deepen mutual trust and understanding with a harmonious spirit.

- Health-First Policy
  We will make a healthy body and mind the basis for personal development.

Figure 1-3 The Four Canon Principles

CANON'S APPROACH TO CORPORATE IMPROVEMENT

CPS was designed to serve as a long-range improvement strategy. But to make fundamental, permanent improvement requires everyone to commit themselves to a process of continual change and growth. Since progress is achieved daily, planning must be thorough enough to support a day-to-day approach. Clearly stated goals are essential, but no end point to improvement must ever be contemplated.

In selecting concrete action for improvement, Canon has always done things in ways that are consistent with Canon's respect for individuality. For example, a year before the Premier Company Plan was announced, President Kaku was impressed by the Toyota Production System. "Let's think about this," he said. "But it must be tailored especially for Canon... Toyota is just one of many superior companies; I want a system that combines the best of each with the Canon tradition."

This pragmatic, individualistic approach is most obvious in the relationship between the different Canon factories. For example, each factory has implemented the HIT system, Canon's version of Toyota's kanban system, in its own way. Camera factories A and B may use "HIT cards" and follow the same rules, but their actual applications can be quite different. (See Chapter 8)

THE CPS THREE-YEAR PLANS

At the outset, three-year medium-range plans were drawn up to promote CPS operations and guide the system's development in every factory.

First Plan (1976-1978)
Build a framework and establish visible control

Second Plan (1979-1981)
Let's do what's been decided

Third Plan (1982-1984)
Achieve higher levels and higher speed

First Plan: Waste Elimination and Visual Control

Corporate improvement began with a company-wide campaign to eliminate waste. Nine wastes were defined (Chapter 2), and personnel at all factories were asked to study questions such as: Where do you see waste; if waste is the difference between reality and the ideal, how can it be expressed as a numerical value; what are the best methods for eliminating waste?

Out of many lively discussions, a CPS textbook and training materials were prepared, and an ideal condition was proposed: since waste is the result of actions or conditions that are not always obvious, systems should be designed that prevent waste from occurring in the first place. For example, defects are a form of waste that can result from inadequate machine maintenance, improper equipment operation, lack of reliable standards or lack of adherence to standards, damage due to excessive handling or inappropriate containers, etc. Eliminating these hidden "wastes" prevents defects from occurring.

Visual control systems were adopted to control existing wastes and to help create a system where waste cannot occur. (The word *visual,* aside from meaning visible to the eye, also means making problems easy to see.) A *visual control* provides a visible standard so that anyone can tell at a glance whether an abnormality has occurred.

Various methods of visual control were studied and introduced. The "Five S" principles of housekeeping and visual control were implemented to achieve orderliness and consistent work arrangements on the line (Chapter 6); visible management systems were also used to promote standardization of manufacturing technology, to improve productivity and work rates, and to help identify and eliminate quality problems every day. The HIT system, for example, is largely a visual system for parts supply and production control. Similarly, Canon's quality assurance flow charts were introduced to visually manage quality assurance procedures and promote greater adherence.

Predictably, these improvement activities did not always proceed smoothly. Through the matrix organization, however, CPS promoted open information exchanges between line sections and between factories. This sharing helped speed up the improvement process and sparked a positive spirit of competition within the company.

The Second Plan: Building Adherence

Stop looking for reasons why it can't be done!
Kaku

As implementation of the first plan proceeded, it became apparent that people tended to blame the plan or the standards first, whenever anything went wrong. For example, when manufacturing defects occurred, people often looked for ways to change the manufacturing method without conducting a thorough investigation into root causes.

Over time, however, workers found that defects usually occurred because someone either did not know the work method that had been established or knew it but did not follow it. It became clear that lack of adherence to established methods and standards was a great obstacle to effective improvement. "Let's do what's been decided" became the slogan for the Second Plan.

Practically speaking, this meant that everyone agreed to adhere to all plans, schedules, and rules. Workers were to follow work guidelines (standard operating procedures). If they could not, they were to inform their supervisor. In this way, they could determine together whether the cause of the problem lay in the standard or in the way the work was performed. The process of improvement became much easier and more efficient.

Focusing on adherence clarifed the responsibility of individual workers and helped them see how their work fit into the overall production process. Focusing on the need for standardization helped Canon develop a system that everyone could understand.

Day-to-day improvement efforts in the workplace were supported during this period by activities in other settings, including the CPS Study Committee, regular meetings to announce results, and other information-sharing activities. *Operation Catchball* and *Doctor's Rounds*, for example, are unique group activities for plant managers and mid-level managers in manufacturing. (See Chapter 5) At the line level, workers and supervisors were also involved in small group activities and projects related to work improvement ideas submitted through Canon's suggestion system. These activities were carried on throughout the company and added impetus to the CPS drive for corporate improvement. (See Chapter 6 for a detailed discussion of Canon's extensive small group activities and suggestion system, the Five S visual housekeeping campaign, and other company-wide information-sharing programs.)

The Third Plan: Increasing Productivity

During the period of the second plan, competition intensified as Canon diversified into areas that made use of its optical and precision technology. This, combined with increasingly shorter product life cycles, increased the need to reduce costs and shorten startup times. Furthermore, a new sales target of one trillion yen ($4.8 billion) by the end of the 1980s meant Canon had to plan for a fivefold increase in sales over the decade.

It became clear that production departments needed a system that could respond much more quickly to these new challenges and to changing market conditions. So, the slogan for the third plan, "Higher levels and higher speed," was most appropriate. Emphasis shifted to a drive for the "ultimate production system" — one that was trouble-free, stockless, and flexible.

Striving to realize this goal involved efforts to increase productivity in every aspect of production, the widespread use of Performance Analysis and Control (PAC). Developed by Takeji Kadota in the late 1960s, PAC increases productivity by improving worker performance. PAC achieves higher levels through careful measurement, self-management, and reporting combined with guidance by front-line supervisors in place of monetary incentives. (See Chapter 8) Other efforts involved value engineering and related cost-reduction activities. (See Chapter 10)

The following chapters look at CPS as a system for managing people, change, and improvement as well as new technology — Canon's "software for success." Chapter 2 explains the central role of waste elimination, a strategy that has resulted in nearly $800 million in savings to date. Chapters 3 and 4 introduce the activities and systems that promote and support improvement at the line-section level — the Premier Work Center drive, factory support offices and organizations, study groups, and information networks.

Effective planning has been a major factor in Canon's success. Chapter 5 explains the methods and programs used to develop individual planning skills at every level, from the plant manager down to the line worker.

Chapter 6 reviews the three principal activities that have successfully promoted employee involvement at Canon — the Five S campaign (visible management and waste elimination at the line level), small group activities for workers and managers, and Canon's work improvement suggestion system. It also describes the communication and awards system that lets Canon employees share and celebrate the results of their efforts.

Chapter 7 describes Canon's training philosophy and methodology, and Chapters 8, 9, and 10 explore in detail the three arms of the Canon Production System — production assurance (PA), achieved through the HIT just-in-time delivery system; quality assurance (QA), emphasizing standardization and the drive to increase process capability; and cost assurance (CA), using value engineering techniques to promote cost-control and cost-reduction activities.

2

Improving the Corporation Through the Elimination of Waste

Canon took immediate measures to promote recovery in 1975, but management recognized that it would take more than a band-aid on surface symptoms to make the kind of fundamental improvement the company needed.

Generally, a downturn in management indicators or a decline in corporate achievement can be traced to internal factors that prevent the company from responding quickly to change. Most of these internal causes are deeply imbedded inefficiencies or "wastes" that go unnoticed because they are buried in everyday operations — in the resources (people and equipment) and in the methods and systems designed to use those resources effectively. *Fundamental improvement* means working to eliminate these wastes that can negatively affect product quality, cost, and delivery time. (Figure 2-1)

Broadly speaking, *waste* is any factor in operations that does not add value. To make these factors more visible, Canon defined nine types of waste:

- Waste caused by work-in-process
- Waste caused by defects
- Waste in equipment
- Waste in expenses
- Waste in indirect labor

- Waste in planning
- Waste in human resources
- Waste in operations
- Waste in startup

Their elimination was made the heart of the CPS drive for fundamental improvement.

Figure 2-1 shows how the nine wastes are attacked from two sides. One approach eliminates waste related to the resources required for manufacturing operations, i.e., labor, equipment, and costs. Another eliminates waste related to the manufactured products, to increase quality and reduce product cost and delivery time.

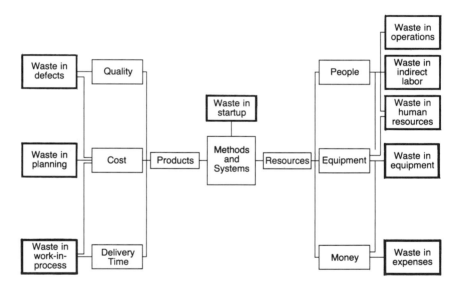

Figure 2-1 Manufacturing Operations and the Nine Wastes

A variety of *management indicators* are established for each type of waste, e.g., work-in-process turnover, reject rate, efficiency rate. These measures are used at different levels within the factory to track and manage improvement. Fundamental improvement is linked to the ability to raise these indicators consistently. Accordingly, results are also evaluated in terms of the nine wastes. There are several advantages to having so many avenues of attack in a company-wide improvement effort:

- Any problem can be approached from several angles, making it easier to see and comprehend.
- The multifaceted approach forces management away from piecemeal improvement toward system-wide improvement.
- The approach is sophisticated; it demands broad knowledge and greater skill. Recognizing this fact challenges workers and heightens their involvement and respect for the process.

NINE TYPES OF WASTE HIDDEN IN EVERYDAY OPERATIONS

Waste Caused by Work-In-Process

This waste occurs in the manufacturing process, between the time materials are brought in and the time they are shipped out as products. Whenever parts stop and wait at different stages in the process, costs that do not add value begin to build up. And, when many components are involved, the problem may become even harder to recognize. For example, space and storage systems for components are accepted as necessities, as well as personnel to manage them. Further, when capital is frozen in inventory, borrowing for current expenditures increases, which adds to the burden of interest. And, when parts or products held in inventory are rendered obsolete by production or design changes, disposal costs are added to the original loss.

Waste Caused by Defects

When the product or part characteristics do not fall within specifications, the waste in materials and labor includes not only the manufacture of the defective item itself, but also rework, disposal, and other indirect costs.

Under the Canon manufacturing control system a part is passed only if its characteristics form a normal distribution curve. Those with abnormal curves are rejected even though they may fall within specifications. (Figure 2-2)

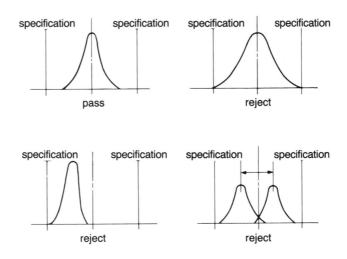

Figure 2-2 How Defect is Defined

Waste in Equipment

Waste occurs whenever physical facilities, equipment investments, or operations are not cost-effective — when poor planning results in the purchase of equipment with greater capacity than required, or when machines and equipment are not operated efficiently.

Inefficient operation can be broken down further into waste caused when equipment is not in operation and waste caused when it is run inefficiently. Measuring progress in equipment operation is based on two different approaches: the operating ratio and the working ratio. (Figure 2-3)

The *operating ratio* measures equipment use — the relationship between machine capacity and the actual time it takes to produce the required amount of acceptable products. This ratio fluctuates as the production schedule responds to changes in demand. The *working ratio* measures the difference

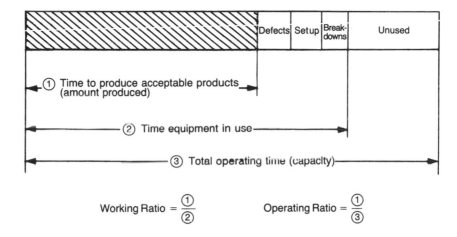

Figure 2-3 Operating Ratio versus Working Ratio

between productive operation (work) and unproductive operation (running with defects, inefficient changeovers, unscheduled stops, etc.). The goal is to approach 100 percent by eliminating waste caused by defects and breakdowns, and by reducing setups.

Waste in Expenses

Expenses include all manufacturing costs in addition to basic materials and outside orders, i.e., all costs related to personnel, supplementary materials, expendable tools, light and heat, etc. Waste occurs in this category whenever the benefits of an item are outweighed by its cost. To help eliminate waste in expenses, they are entered on a zero basis at the budgeting stage and then brought within the budget figure at the stage of actual operations through cost-reduction team activity. (See Chapter 10 for more on cost-reduction and cost-control activities.)

Waste in Indirect Labor

This form of waste can be found in indirect operations and personnel allocations that do not add value. This occurs whenever management systems are poorly planned or when management functions are poorly exercised.

Waste in Planning

Canon looks for waste in planning in two areas — in manufacturing process and purchasing strategy. Waste in manufacturing process occurs when inappropriate manufacturing methods, machinery, or materials handling make costs for a particular function too high. Waste in purchasing strategy occurs when purchases are made without orderly purchasing procedures or adequate study of market trends or feasibility.

Waste in Human Resources

Human resources are wasted whenever individual abilities are not used fully, i.e., when people are given work that could be done by machines, rather than tasks that require human judgment and creativity.

Waste in Operations

Waste in motion occurs on the production line when standard procedures are either not followed or not carried out efficiently. It also occurs when the standard procedures themselves are inefficient.

Waste in Startup

More effective planning can reduce wasted startup costs for new products. Excessive startup time — from trial mass production to stabilization — can be reduced by more thorough technological study at the trial mass production stage. Because waste in startup is often a combination of wastes in planning, defects, and motion, etc., it is an excellent general measure of manufacturing efficiency.

MEASURING RESULTS OF WASTE ELIMINATION
AND OVERALL IMPROVEMENT

Fundamental corporate improvement is measured in terms of waste elimination results and expressed as a profit. Expressing improvement as a profit has several advantages. Everyone involved becomes more conscious of profit and loss, and a lively, positive competition develops between factories and divisions. Finally, overall improvement can be evaluated easily by combining the results of different waste elimination efforts.

Improvement resulting from waste elimination is calculated by the following method:

Waste Elimination Profit (WEP) = *the degree to which management indicators have improved over the previous year's record, expressed as a dollar amount.*

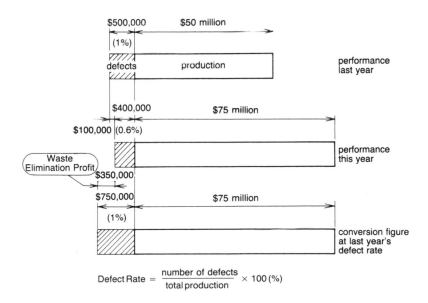

$$\text{Defect Rate} = \frac{\text{number of defects}}{\text{total production}} \times 100\,(\%)$$

Waste Elimination Profit = (prior [defect] rate − current rate) × current production

Figure 2-4 Waste Elimination Profit from Defects

An example of this calculation using waste from defects is shown in Figure 2-4. The actual difference between this year's and the previous year's cost of defects is $100,000. Canon looks at it this way, however: if this year's rate had remained the same as last year's, then the actual cost of defects would have been $750,000, or 1 percent of $75 million. But, since waste elimination efforts reduced that to $400,000, there is a *waste elimination profit* or WEP of $350,000.

The Nine-Waste Elimination Profit is the total profit figure from all nine types of waste, each calculated in the same way. Since this figure differs according to the scale of individual factories, however, it cannot be used for comparative purposes. To express the degree of overall improvement for any particular factory, the total figure is divided by the number of personnel to give WEP per person. This calculation is done at every level — section, division, factory, and company. (See Figure 2-5)

RAISING FACTORY PROFITS THROUGH WASTE ELIMINATION

Figure 2-5 shows the company-wide increases in WEP at Canon and affiliated companies from the beginning of CPS to the present. Note that although WEP indicates improvement over the previous year, a drop in WEP does not indicate deterioration until it becomes a minus figure.

WEP expresses the benefit of overall improvement in dollars, so it differs from the after-cost profit figure used in financial accounting. When WEP went over $50,000, however, top management at Canon began to take interest in the relationship between WEP and financial accounting.

Factories are required to produce a profit plan for the product groups as well as a waste elimination plan for CPS. Typically, these profit plans are designed for 100 percent achievement, so goals are set at moderate levels. Waste elimination plans, on the other hand, are for internal management use, so very challenging goals are demanded. *In practice, actual oper-*

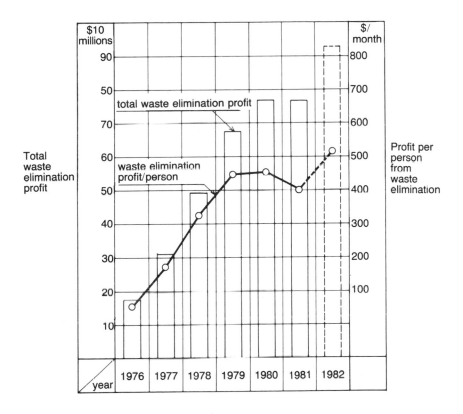

Figure 2-5 Total Nine-Waste Elimination Profit (Including Affiliates)

ations are developed around the waste elimination plans, because reaching the higher WEP targets will also result in achievement of the lower profit plan targets.

The points in common between the CPS waste elimination profit and the after-cost financial profit are factors having an influence on profit such as the work-in-process turnover and defect rates. There is only one goal-achievement plan to raise the level of these indicators. What differs is simply the method of calculating profit.

The relationship between waste elimination efforts and traditional profit planning becomes even clearer when the results of waste elimination are examined closely. In every case, eliminating waste either increases value added or reduces fixed costs:

1. *Eliminating waste caused by work-in-process* means reducing costs for warehousing, facilities, and management personnel.
2. *Eliminating waste caused by defects* means reducing variable costs and increasing value added through greater efficiency.
3. *Eliminating waste in equipment* means improving the working ratio, increasing production volume through decreased cycle time, and reducing equipment expenditures.
4. *Eliminating waste in expenses* means reducing fixed expenses.
5. *Eliminating waste in indirect labor* means reducing labor costs through reductions in indirect personnel.
6. *Eliminating waste in planning* means adding value through value engineering and cost reduction.
7. *Eliminating waste in human resources* means reducing labor costs by introducing labor-saving equipment.
8. *Eliminating waste in operations* means adding value through increased efficiency as well as increased production capacity.
9. *Eliminating waste in startup* means reducing setup for equipment and processes as well as reducing trial mass production costs.

By increasing value added or reducing fixed costs, waste elimination efforts work to improve the break-even point ratio. (Figure 2-6) This is the point of contact between CPS and financial accounting as well as the point of departure for CPS activities. Although overall improvement in operations is the immediate aim of CPS, its ultimate goal is the profitability that can support the company's continuing vitality and growth.

THE FOUR INVESTMENTS

The CPS waste elimination program is designed to reduce waste *now*, through front-line efforts involving personnel on all

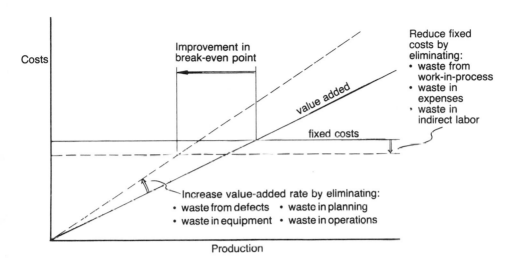

Figure 2-6 Waste Elimination Effort and the Break-Even Point

levels. Planning a company of the future that can respond quickly to change, where waste cannot occur, is a strategy involving the top levels of factory management — the *Four Investments*.

- New technology
- Developing human resources
- Equipment
- Human welfare

The four investments concept was developed to meet the resource needs identified through waste elimination activities and to plan for systematic, sensitive reinvestment of waste elimination profits.

Currently, investments in new technology have the highest priority and include research and development in high-precision manufacturing technology, implementation of new techniques to simplify the use of new materials, and so forth. This is an important outlay if Canon is to continue developing "products that make a difference." Equipment expenditure is also important in determining the shape of manufacturing in the near future. To use equipment that provides the most flexibility, it is essential to identify trends in both hardware and software technology.

Investments in human resources include on-the-job training and other educational and training programs. Since spending in this area rarely produces an immediate effect, it is all the more important to look to the future, plan, and take action early. Investments in human welfare are intended to realize Canon's commitment to health. The goal is to build a workplace environment that emphasizes people — one that is comfortable, stimulating, and easy to work in.

Keys to Fundamental Improvement

Understanding and eliminating waste is the key strategy for permanent corporate improvement.

By the end of 1983 cumulative waste elimination profits at Canon totaled over $450 million. Today, waste elimination efforts are focused primarily on indirect operations rather than manufacturing, but the savings continue. Waste elimination is the theory; CPS is the practice. Chapters 3 through 10 examine in detail how waste elimination consciousness has helped Canon make fundamental frontline improvement (Chapter 3), direct and manage the improvement process (Chapter 4), focus day-to-day target setting and goal achievement (Chapter 5), strengthen worker involvement and training programs (Chapters 6 and 7), and implement a highly efficient production system. (Chapters 8-10)

3

Building the Premier
Work Center

STRUCTURE OF THE CANON PRODUCTION
SYSTEM (CPS)

Canon's goal under the Premier Company Plan is to achieve the best quality, lowest cost, and fastest delivery anywhere — through reliable quality assurance, the ultimate continuous flow system, and comprehensive development of human resources.

Waste elimination is the principal CPS strategy to support this growth; it determines how problems are identified and defined and how the results of improvement are calculated. And, it produces continuous improvement (and profits) in every aspect of the company — equipment, personnel, design, management, and others. However, the manufacturing line section or work center is the starting point for all CPS improvement activities; building the ideal or premier work center is its ultimate goal.

Figure 3-1 shows the three systems or "pillars" created to support this improvement effort: 1) the line-centered *basic production system*, concerned with quality, cost, and delivery time; 2) the staff-centered *support system*, supporting the line effort with production technology and management techniques; and 3) the *workplace vitalization* program, promoting continuous im-

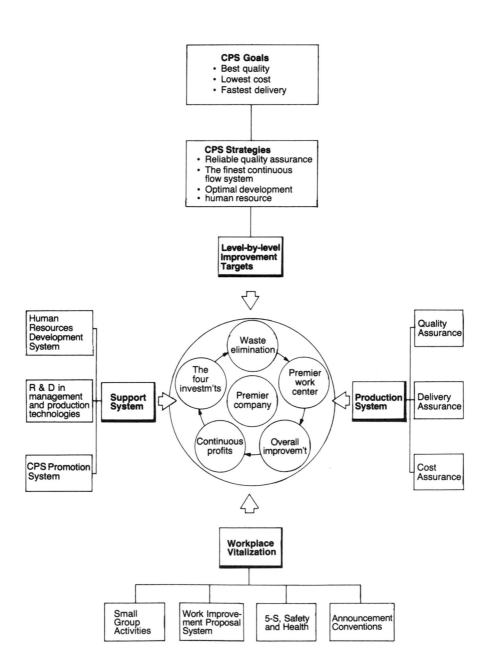

Figure 3-1 Basic Organization of CPS

provement through structured group involvement and activities.

To achieve results efficiently, *level-by-level targets* assign functions and goals to each level in the organization. Then the profits resulting from the overall waste elimination effort are reinvested in the Four Investments to strengthen the company for the future.

The Basic Production System

Three separate systems assure optimal quality, delivery time, and cost. In the quality assurance system, QA flow charts, standardization, process capability, and quality improvement activities help build in quality at every stage of the manufacturing process. The delivery-time assurance system uses the principles of kanban to achieve shorter manufacturing times and quicker responses to change — the ultimate continuous flow system. And, the cost assurance system reduces and controls costs through value engineering team activities, through effective management of purchasing and subcontracts as well as quality and productivity improvement on the production line.

The Support System

The support system illustrated in Figure 3-1 operates at every level — in company headquarters and in every factory. The human resources development system, discussed later in Chapter 7, provides training in management and technical specialities, skill broadening programs, and training for overseas personnel. Management and production technologies to meet Canon's needs are developed, introduced, and supported by a variety of special and company-wide projects. Finally, the CPS promotion system organizes CPS activities at the company and factory level, arranges special functions, and maintains the awards system. (See Chapter 6)

Workplace Vitalization

Getting everyone involved and giving their best effort is essential to the improvement process. Canon production operations had been periodically vitalized in the past by the zero defect and workplace cleanup movements and by meetings to share improvement techniques and results. Under CPS, these activities are now organized and managed separately as: 1) small group activities, 2) the Five S campaign, 3) the work improvement proposal system, and 4) announcement conventions. Each of these activities is discussed in detail in Chapter 6.

CREATING THE PREMIER WORK CENTER: CPS IDEALS IN ACTION

CPS goals are actually achieved through production in the smallest operating unit—the single line section or *work center*. For this reason, all CPS improvement activities within the sections are aimed at achieving the most efficient and productive operation—*the premier work center*. (See Figure 3-2)

```
──────────────── In a Premier Work Center: ────────────────

 • Achievement is constantly high.
 • People are growing.
 • Operations are always being improved to eliminate waste.
 • Individual functions and goals are clear.
 • Everyone sticks to what's been decided (adherence
   and follow-through).
 • Anyone can see how things stand (visual control).
```

Figure 3-2 Premier Work Center Profile

To create a premier work center three conditions must be present. First, the work center environment and the improvement process itself must provide continuous opportunities for growth to every manager, supervisor, worker, and support staff member. This means sharing information and providing everyone with opportunities for meaningful participation. Next, everyone must understand what is required to transform the section. This means defining the ideal and the actual situation in concrete terms, identifying the gap between the two, then carefully and cooperatively planning steps to bridge that gap. (Figure 3-3, steps 1-5) Third, a plan must be developed and implemented in a manner that ensures achievement of CPS goals. A successful plan requires everyone's cooperation and

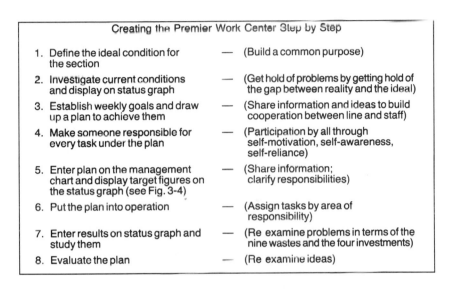

Figure 3-3 Creating the Premier Work Center Step by Step

willingness to "do what's been decided" — to be observant and keep track of progress and to reevaluate and adjust the plan, on a day-to-day basis if necessary. (Figure 3-3, Steps 6-8) In order to keep everyone informed — an important aspect of visual control — goal achievement plans and results are displayed in premier work center management charts and work center achievement graphs. (See Figure 3-4 and Photo 3-1)

Gaining everyone's cooperation in the improvement effort can be difficult when communication between levels is slow or restricted. Before the creation of CPS there were seven levels between the factory manager and workers, four alone between workers and the section chief. This stratification produced misunderstandings and mistrust between levels and worker apathy.

CPS Premier Workplace Management Chart

Year: _____

Category	Area of concern	Main points	Person in charge	Deadline		Yr: ___ actual figure	Yr: ___ target figure	Plan and results by month											
				date scheduled	date completed			1	2	3	4	5	6	7	8	9	10	11	12

Figure 3-4 Premier Work Center Management Chart

To overcome these problems, each factory section was reorganized in 1978 into blocks of 10-20 workers managed by a

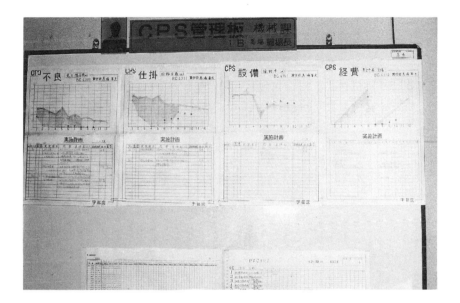

Photo 3-1 Work Center Achievement Graphs

single foreman who reported directly to the section chief. (See Figure 3-5)

The block system has contributed greatly to the CPS premier work center goal. For example, it gives primary responsibility for supervision and guidance to one person — the foreman. Since effective supervision is one of the basic principles of PAC (Performance Analysis and Control), that program has functioned more effectively as a productivity control system since the change. Bringing workers, foremen, and section chiefs closer together has improved communication, participation, and teamwork, and increased overall productivity.

Case Study 3.1 — Aiming for the Premier Work Center at Factory F

Before CPS was introduced at Factory F, each section set

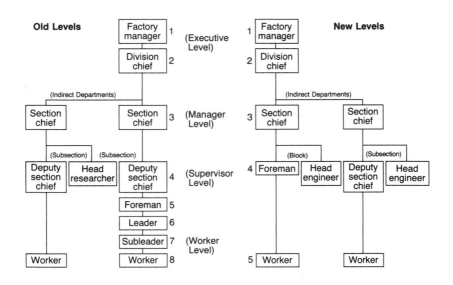

Figure 3-5 New Factory Organization

goals independently. Typically, it took all their efforts simply to achieve the current year's goals. However, the CPS principles of waste elimination gave factory management a common measure to use in evaluating different section operations and in planning long-range improvement. Quality and productivity began to improve dramatically as section chiefs concentrated on the nine wastes in planning and foremen attacked excess work-in-process, defects, and waste in motion on a daily basis.

In 1979, Manufacturing Section 2 had 4.3 days worth of work-in-process, defects at 0.94 percent, and a 92 percent performance rate. At that time, the section chief announced his intention to win the premier work center prize — the highest honor a section could receive — by 1981.

He began by initiating an educational drive to make sure all workers, foremen, and staff members thoroughly understood the principles of waste elimination and their own role within it. Newly hired workers were briefed exhaustively; workers who did not understand fully came back to study on their days off. The foremen became determined to win the prize, "no matter what!," and their commitment began to affect the workers.

More and more workers spoke up in CPS study meetings to express their opinions or ask pointed questions. Groups of workers stayed after work to engage in heated discussions about plans to improve the machinery.

In this manner, CPS principles were passed from one level to the next until everyone was comfortable with them. For example, in the past when control limits had been exceeded, some foremen had allowed work to proceed without taking corrective action. Often no action was taken until defects began to show up in subsequent processes. Now, statistical methods were introduced to handle internal data for the section. And at every work station waste elimination control charts were displayed and carefully maintained. Whenever abnormal figures appeared, their causes were promptly investigated and indicated on the graphs.

Keeping promises became an important daily objective. If this seems like an overly simple approach, consider what happens when parts or tools are not ready on the date promised, when defects occur because established standards are not adhered to, or when documents are not produced on time. These failures are common and predictable, but the cumulative loss they create in any given period is considerable. Clearly, if all levels consistently deliver on their promises, the corporation gains an unbeatable competitive edge.

To achieve this, many sections started a promise-keeping drive. In one section it was called KSY Drive (from the Japanese words meaning "Honor commitments immediately"). Another section had its KKJ Drive ("Follow through on promises every time").

In Manufacturing Section 2, this drive was called the *PDS Drive* (from the English words "plan-do-see"). It was given this name because the ability to keep promises depends on careful planning and frequent plan adjustment based on evaluation of actual results. Everyone in the section kept a PDS notebook, to keep track of assigned tasks and schedules and make sure they were completely carried out. And to reinforce the group effort, a promise-keeping status chart on everyone was posted behind

Candidate: Mfg. Section 2 **Representative:** _____	**Number of Personnel:** direct 92, indirect 14 (as of Dec. 8)

Situation Through Last Year

• This section (processing parts for steel camera main bodies) has been kept jumping by annual production increases and new plans (new products, SGS coating, optical MO, etc.).

• In 1979 we were runner-up Premier Work Center. However, this wasn't as much from waste elimination as from the tie-in between production increases, new plans and the increased productivity of CPS. We didn't know how to eliminate waste (increase efficiency) without increasing production at the same time.

• With over 200 people, some details were overlooked. But in 1980, the old Sect. 2 was split and both workers and work systems took on a new life.

Goals Emphasized this Year	**Concrete Steps:**
We are all determined to become Premier Work Center this year for sure! 1. PDS Campaign — A method for pushing through even complicated improvements reliably and on schedule, to produce a planned effect on OUTPUT.	Plan Promotion: Annual, 6 month and 3 month plans. 1. Break down plan in detail, itemize things to be done, person responsible and schedule – one point per sheet. Use PDS Plan Do See sheets to prevent omissions. 2. Review and correct target figures in case of production changes, stoppages or new plans. 3. Eliminate gaps in all types of planning sheets at monthly CPS study meetings.
2. Total Participation in CPS — Get everybody to put their heads together and give their ideas! (Lots of small improvements = big results) — increase everyone's understanding and cooperation.	To Improve Participation: 1. Monthly CPS block meetings (foreman and workers) • Schedule after hours or during rest periods so everyone can attend. • Set schedule first of the month. • Small groups study assigned topics, share ideas and review plan results. • Staff from factory office and section chief participate as observers to encourage and support participation and follow-up. 2. Bi-Monthly CPS increase rallies (foremen, leaders and staff) • Draw up plans • Follow-up problems that can't be solved within the block.
3. Visual Control — Make progress toward plan goals easy to see — make all problems on the line visible at-a-glance.	1. Make Plan and Progress Visible. • All plans use 5W1H (who, what, when, where, why, how) • All gaps are filled in as implementation is completed • All changes are reviewed and adjusted immediately • All changes are indicated visually so they can be seen at-a-glance, at any time. 2. Organize the line to make problems visible. • Live work-center management — gather data linked with action constantly (quality, PM, etc.) • Establish standards, i.e., a place for everything • Make defects visible: Daily handling, self management, problem management through linestops • Make efficiency rates visible: Make work plans to show rate from hour to hour

Figure 3-6 Application for Premier Work Center Status

3–Year Record (yearly average)			
Item Year	'79	'80	'81
Work-in-process rotation	4.3	2.49	1.58
Defect rate	0.94%	0.382%	0.172%
Overall performance	92.34%	98.47%	100.4%
Improvement rate	12.6%	5.8%	14.3%
Proposals submitted per capita	15.2	25.6	45.6
Work attendance rate	96.5%	96.6%	97.1%

Effect (Present Conditions - Outlook for Future)

(Present Conditions)

• The present system helps implement plans successfully and ensures that problems identified during implementation are included and dealt with under the next plan. It also facilitates speedy transfer of improvements to other work centers.

• Under the present system, all workers are involved in CPS activities and the target indices for small group activities are directly linked with CPS goals.

• Plan goals and current status can be compared from moment to moment. All action taken is based on data.

('81 Results)

	Standard value	Target value	Effect	Achievement rate
Motion	82.0%	92.8%	100.4%	170.4%
Work-in-process	20,050 pieces/B	13,300 pieces/B	12,500 pieces/B	111.9%
Defects	0.332%	0.191%	0.172%	110.0%
Improvement plans		30 proposals per person/yr.	45.6 proposals per person/yr.	152.0%
Value added	$117.40	$136.02	$150.95	180.1%

(Outlook for Future)

1. Increase achievement levels for the present system.
   - Refine PDS method.
   - Standardize operations involving intuition and know-how.

2. Improve Preventive Maintenance (PM).
   - Stabilize and maintain present improvements, promote operator training.

3. Build system that responds to change.
   - Maintain work plans, aiming for greater speed.
   - Improve technological understanding.

Figure 3-6 (cont.)

the section chief's desk where it could be seen easily. A circle indicated a promise kept, an X a promise broken. At first, X's averaged 20-30 percent, but gradually these were buried completely by circles.

Figure 3-6 shows the improvement in all management indicators resulting from the section's efforts. Changes in the workplace itself were also striking. Cleanliness became such an important focus of worker improvement activity that visitors often asked how a section using oil could be kept so clean.

Many valuable improvements were introduced:

- The productivity and process capability of certain equipment were increased by raising standards for operators, improving the preventive maintenance system, and taking measures to increase blade life in cutting machinery.
- A system of half-day rotation for hand-assembly line work increased worker versatility.
- A "morning pickup" system was introduced to reduce work-in-process to one day in multi-model, small-lot hand manufacturing operations. (This drive to ensure completion of each day's scheduled production, i.e., the "morning pickup," is explained in detail in Chapter 8.)

As a result of its continous efforts in these and other areas, Manufacturing Section 2 became the proud recipient of the Premier Work Center prize in 1981.

Keys to Fundamental Improvement

Concentrating improvement efforts in the smallest production unit yields the highest results.

The work center is the place where productivity and quality challenges must be met. For this reason, it is the focus of all improvement activity under the Canon Production System. When every section strives to eliminate waste and to achieve the most efficient operating conditions, the entire company benefits.

A company-wide commitment to continuous improvement cannot be maintained, however, without effective administrative support. Chapter 4 describes the basic CPS support organizations and some of the communication networks and information-sharing activities that promote day-to-day improvement in every factory throughout the company.

4

Ensuring Continuous Improvement: Canon Production System Administration

MANAGING IMPROVEMENT — BASIC PRINCIPLES

The purpose of CPS is to improve and strengthen manufacturing operations by concentrating on goal achievement and improvement at the line level. To realize these goals, CPS administration provides the structure for a simple, orderly process of planned growth. First, target figures are set for waste elimination profit and for the various management indicators (defect rate, productivity, etc.). Then to achieve these targets, day-to-day implementation or goal achievement plans are drawn up and put into action. To keep this process functioning successfully in a company as large as Canon requires effective management and support. Each factory manages its improvement process a little differently, but all follow the same basic principles.

Focus on Overall, Continuous Improvement

Total waste elimination profit is expressed in dollars in planning and announcing factory results. When targets for management indicators are issued to the factories by CPS, however, they appear as a recommended rate of increase or decrease over the previous year rather than an absolute figure. In this way, improvement is always viewed as a continuous process.

Emphasis on Planning

Each factory draws up an annual goal achievement plan, taking into account the recommended CPS indicator targets and other factors such as the objectives emphasized in the three-year plan and product group policies. This goal achievement plan serves as the basis for individual plans drawn up at every level within the factory.

Planning Involves Everyone

Planning starts in the smallest units with self-management and small group improvement activities that stress the plan-do-see cycle. Planning (based on local problem identification and data collection) moves up from level to level (block → section → division → factory). Assignment of goals moves downward (factory → division → section → block). Planned goals are evaluated and adjusted at each level in terms of their practicality, urgency, and balance with other concerns.

Sharing Information Vitalizes the Workplace

Outsiders are sometimes surprised at the number and variety of organized meetings and announcement conventions (rallies to share goals and achievements) sponsored by CPS. However, these events promote a healthy, open atmosphere; individuals learn to share information and expertise with each other, easily and efficiently, in order to achieve common goals.

CPS Builds on Improvement

Many improvement measures adopted as part of the CPS effort had been previously developed and implemented by the existing organization. For example, the block system and the systematization of quality, cost, and delivery assurance functions provided a solid foundation for CPS within manufacturing operations.

This chapter focuses on the CPS organizations and management activities that support the system, particularly those that promote effective communication and information sharing. Goal setting and planning are described in Chapter 5, workplace vitalization and announcement meetings in Chapter 6.

CPS ADMINISTRATION

The Canon Production System is promoted through:

- *support organizations* operating within the existing company and factory structure;
- *group discussion and planning*, i.e., study groups, committees, conferences, or project teams; and
- *special upper management programs* for the exchange of information and techniques.

CPS Support Organizations

As the CPS drive built up momentum on the production lines, permanent staff support organizations were set up at company headquarters and within the factories. (See Figure 4-1)

Factory organizations. Factory support systems are provided to strengthen QCD assurance. The Quality Assurance section establishes product quality standards, designs QA flow charts, and is responsible for measurement control and shipment assurance. The Cost Assurance office is responsible for manufacturing cost-planning and promotion of value engineering. The Production Assurance section establishes standard production figures and startup schedules for new products, and promotes the HIT production system and improved materials flow.

In every factory there is also a CPS office devoted exclusively to promotion activities. These offices gather and process the results of goal achievement and other improvement efforts and sponsor planning and announcement meetings; they exchange waste elimination techniques with other factories, promote various educational activities, and serve as factory liaison with the main CPS office.

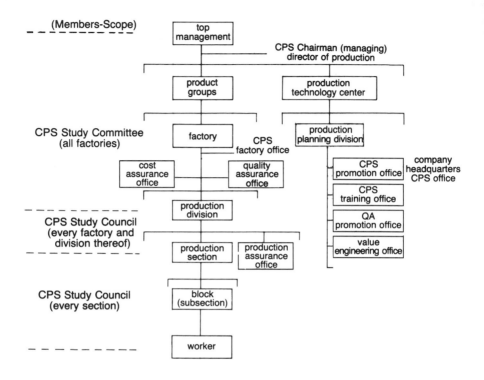

Figure 4-1 CPS Support Organizations

Headquarters organizations. At company headquarters several important offices are attached to the Production Technology Center and the Production Planning Division.

The CPS Promotion Office includes central offices for the work improvement proposal system and small group drives as well as for CPS. It conducts factory evaluation and industrial engineering studies to help improve production and management systems. The CPS Training Office plans, develops, and promotes training within the manufacturing division.

The QA Promotion Office is responsible for QA systems and methods and for setting manufacturing technology standards. The Value Engineering Promotion Office is responsible for standardizing cost estimates, for cost control, and for promotion of value engineering projects.

CPS Group Discussion and Planning

Company-wide CPS Study Committee. The CPS Study Committee is the highest decision-making body within the Canon Production System. Its members include the CPS chairman (managing director for production), the director of the Quality Assurance and Production Technology Centers at company headquarters, all factory managers and manufacturing division chiefs, top representatives of all affiliated companies, and other managers involved in planning at Canon and its affiliates.

The committee studies the three-year promotion plans, establishes annual company-wide targets for waste elimination profit, approves annual promotion plans, issues progress reports, sponsors the presentation of awards, and selects the winners.

In addition to generally coordinating CPS efforts, the committee plays a significant role in disseminating improvement techniques. Meetings are held at a different factory each month so that committee members can see the results of the factory's waste elimination effort and share ideas for improvement.

Factory CPS study councils. Factory CPS offices sponsor separate study councils that meet monthly at the factory, division, section, and block levels. The division study council, for example, includes the division chief, line and staff section chiefs, and the individuals responsible for promoting elimination of each type of waste. With the help of staff departments, these study groups track the progress of annual goal achievement plans and research measures to eliminate obstacles.

The Nine-Waste Elimination Promotion Team. Personnel from every factory make this team a cross section of the entire company. Its activities are organized by type of waste, so representatives of every aspect of production, development, and indirect support operations are included. The team's chief functions are to analyze waste elimination results and promote the development or transfer of effective techniques and methods.

Individuals in every factory and in each division are responsible for promoting elimination of each type of waste and a central office for each waste is maintained at company head-quarters. Taking defects as an example, the chief of each factory QA section is a member of the team, and the central office on waste caused by defects is administered by the director of the Quality Assurance Center.

To support waste elimination efforts, the team promotes the development and exchange of waste elimination techniques from factory to factory and section to section.

The team also conducts application studies. For example, in the factory as a whole and in individual work centers, all types of waste are confronted together. As a result, when the overall elimination score improves in one area, group members sometimes close their eyes to the areas that did not improve. However, the waste elimination promotion team investigates each area of waste independently to identify why a score did or did not not improve, and what can be done about it. The team has been particularly effective in operations using fixed tools where no progress has been made. It has also been helpful in situations where quick results are sought, simply by making available ideas and methods already in use at other factories. This team approach to localized problems, organized by type of waste, cuts across all line waste elimination activities and has resulted in increased overall improvement.

In some cases highly specialized teams are formed to deal with modifications that may substantially change the work operation or projects that will require a lengthy development period. For example, the Production Assurance Promotion Committee, which handles the HIT system, was formed to deal with waste in work-in-process, and the Productive Maintenance Liaison Committee was formed to deal with waste in equipment. The Production Preparation Study Committee was formed to deal with waste in startup. The latter committee is sponsored by the director of the Production Technology Center, and its members are division chiefs from every factory. Its chief function is to conduct studies on the startup system for new products.

Upper Management Information Exchange

When it comes to improvement, each factory has its own manufacturing operations and unique areas of concern. The effort invested in improvement can also vary depending on the seriousness of problem addressed, the methods used to overcome it, and the level of worker involvement. Factory managers need individualized diagnosis and guidance for their operational problems; they also need a way to share their ideas, experiences, and successful techniques with other managers who are dealing with the same kinds of problems. Two programs respond to these needs: Doctor's Rounds — personal visits from the CPS chairman, and Operation Catchball — discussion forums for section chiefs who run similar operations.

Doctor's Rounds. Managing Director Yamagata (CPS chairman and managing director for production) makes a "house call" to each factory on an average of once a month. (Photo 4-1) At that time, the factory manager and division chiefs report on the progress of the improvement effort — the goal achievement plans, current CPS profits, and countermeasures used for specific problems. Mr. Yamagata then offers comments and suggestions. He draws on his familiarity with the achievements of other factories and on his own extensive experience as a factory manager and president of an affiliated company. He may comment on a proposed solution, suggest new approaches in an area where little progress has been made, or point out a problem that has not been noticed by factory management.

Chairman Yamagata's visits are invariably stimulating and challenging, not only by reason of his status in the company, but also because of his extensive and long experience. As he walks through the production line, he examines every plan, progress, and achievement chart minutely. He does not hesitate to point out problems, although his way of doing so may force people to stop and think. He may say, for instance, "This plan of yours is a lifetime plan." If the person addressed thinks Mr. Yamagata means the plan is good enough to be used for a lifetime, he is greatly mistaken. What he has really been told is that a plan that

Photo 4-1 Managing Director Yamagata

Photo 4-2 Production Technology Center Director Naitoh

is too abstract to get at root causes can be used for a lifetime without solving problems. However, there is never any need for the person criticized to feel discouraged. The managing director does not assign blame. Just as waste elimination is measured in terms of profit or improvement, an individual's will to succeed and the actual progress he makes are always emphasized.

Operation Catchball. The Doctor's Rounds provide a formal setting for talks with factory managers and division chiefs; Operation Catchball brings together section chiefs to play "catch" with ideas and problems. Section chiefs running similar operations such as large-lot assembly or exterior finishing meet together with Production Technology Center Director Naitoh (Photo 4-2) to discuss related technology and management methods.

Director Naitoh uses these forums to gather information on current conditions and concerns and to inform participants of new directions being considered for the production system.

The main purpose of the forums, however, is to permit the exchange of information and problem-solving techniques among section chiefs. Participants present concrete examples of problems they have been unable to solve; sometimes they visit a group member's workplace and discuss problems on the spot. When a solution cannot be found, a team is formed to study the problem further. Sometimes the discussion breaks down and participants end up consoling each other over the enduring problems no one outside their area can appreciate. Since these groups are made up of cross sections of specialists, however, their discussion can often go beyond the internal circumstances of individual factories and result in company-wide improvements.

Before the introduction of Operation Catchball some horizontal exchange occurred between people at the same level in different factories, but this was largely confined to section chiefs from indirect departments such as product technology, production management, and testing. Since there had been virtually no contact between manufacturing section chiefs, Operation Catchball has been very well received. At present there are twelve groups of related section chiefs meeting once every two or three months.

Keys to Fundamental Improvement

Managing improvement means sharing information.

Improvement means growth and change — two things that cannot occur in a manufacturing environment without thorough, continuous communication and information sharing at all levels and across departments.

Communication must be open and blame-free to break down resistance to change, and the information shared must be concrete, objective, and sufficiently detailed to promote growth. This chapter identifies the productive communication patterns at Canon that have been formalized and reinforced by CPS organizations and activities. Chapter 5 looks at the content of that communication — by outlining Canon's model for company-wide planning and implementation.

5

Target Setting and Goal Achievement: Planning at Every Level

A BUILT-IN EMPHASIS ON PLANNING

"Plan first, then act" is the general rule for all CPS operations. Each is supported by a broad three-year plan and an annual goal achievement plan for day-to-day implementation. The annual plans have become models of precision and effectiveness, because CPS set out to make planning an automatic activity and to achieve two important objectives: greater precision in planning, and greater reliability in plan execution.

To achieve the first objective, all employees are taught the importance of planning in CPS and its basic principles, and they are given plenty of opportunity to sharpen their skills through actual practice, repeating over and over the pattern of *plan* → *execute* → *plan* → *execute*. This chapter describes the planning process and practical activities that make reliable planning a natural part of every Canon employee's day.

When any plan is put into action, discrepancies, or "gaps," may appear between actual and targeted figures. In some cases these differences are caused by changes in conditions since the plan was drawn up; others are caused by inadequate gap management. To manage the gap effectively and make planning more reliable, discrepancies must be adjusted or eliminated as they appear. In every case, someone is immediately assigned to identify the source of the problem and rectify it. This controls the day-to-day variation and helps ensure a perfect correspondence between final plan and result figures, i.e., *Plan = Results.*

Guidelines for Planning

1. The purpose of planning is to help us reach our targets.
2. Before developing a plan, identify the concrete steps to be taken and document the feasibility of every proposed improvement.
3. Before implementing an improvement, estimate and evaluate the expected results.
4. Keep everyone informed through visual control. This means every final plan must be formally announced and posted.

The CPS campaign to improve planning skills is promoted in several ways. First, and most important, people support each other's efforts by sharing their ideas and experiences in planning. Second, planning activity is constantly emphasized. Summaries of the annual achievement plans and their results are announced at company and factory meetings. This information is displayed prominently in the premier work center management schedules and status charts posted in every work area. Third, the campaign is actively supported by CPS management. The chairman of the CPS Study Committee gives advice on proposed plans and tips on gap management through the Central Nine-Waste Elimination Committee and the Doctor's Rounds program described in the last chapter. Finally, planning is worked into everyone's daily routine through a variety of struc-

tured activities that are essential to the continuous improvement process — level-by-level improvement targets, goal
achievement planning, and the self-management drive.

LEVEL-BY-LEVEL IMPROVEMENT TARGETS

At each level in the organization, management planning
functions and goals differ, depending on the scope of responsibility and the time frame. Figure 5-1 outlines the concrete
targets for different levels. Factory managers and division chiefs
plan broad improvements and innovations in factory facilities
and systems. Section chiefs and foremen are expected to increase productivity through nine-waste elimination efforts
focused on the 5Ms — machinery, material, manpower,
methods, and measurement. Foremen, in particular, work to
improve worker performance, quality, and the work environment. Workers focus on work improvement, following the standard, "Fix rules for your own work and follow them."

Target figures for division and section chiefs are based on
the current three-year plan and achieved through successful implementation of the annual plan. Foremen's targets, on the
other hand, are the combined figures assigned to their workers.

Level	Time Frame	Amount	Objectives	Nickname
Factory manager and division chiefs	3 to 5 years	System renewal every 5 years	Systems, equipment, people	"5-year New"
Section chief	1 to 3 years	Reduce waste by 50% in 1 year	5 M's *	"1-year Half"
Foreman	1 month to 1 year	3% increase in 1 month	Increased productivity	"Three-a-month"

Figure 5-1 Level-by-Level Improvement Targets

FROM GOAL SETTING TO GOAL ACHIEVEMENT

CPS improvement targets (projected figures for waste elimination profit) begin as guidelines presented to the factories. These guidelines reflect the major management indicator targets from the current three-year plan and actual results for the past year. For example, the plan may specify a 40 percent decrease in defects, a 30 percent decrease in work-in-process rotation time, and a 30 percent increase in productivity. Based on these figures and the annual profit planning goals of the product groups, factory management then establishes its own indicator targets and calculates the factory WEP targets.

The factory targets are sent down to be recalculated level by level, from division to section and from section to block. Finally, workers' goals are established for the measures pertaining to their own work, i.e., defect or efficiency rate.

Once goals have been set at every level, planning for goal achievement begins in each work section and involves everyone — section chief, foremen, and workers. The first step in the planning process is to select areas to be targeted for improvement.

Problem identification is a continuous process in every work section; generally, the foremen and section chief will have documented 100 or more opportunities for improvement within the section, considering both hardware and software issues. The task in planning, however, is to calculate where improvement is likely to be greatest. The problems ultimately selected are then worked into the plan on the basis of their urgency, the difficulty of solution, and the expected benefit. See Case Studies 5.1, 5.2, and 5.3 for examples of goal-setting and planning at individual factories. [Editor's note: Actual changes in operations, equipment, layout, work methods, etc., are the "hardware" of improvement, e.g., low-cost automation or small dedicated equipment. Managerial factors — ways to promote, measure, and maintain the rate of growth — are the improvement "software," e.g., the level-by-level goal achievement planning process.]

Typically, a factory's annual goal achievement plan includes the detailed improvement plans developed at each level by division chiefs, section chiefs, and foremen. It will also include plans to *promote* improvement — announcement meetings, displays for individual and team results, and other activities to encourage involvement. These promotional plans are usually developed by the section chief and the foremen.

Throughout the year, monthly division meetings and weekly section meetings are held to report on progress and study problems that arise. At and between meetings, intensive gap management is carried out to fill any gaps between target figures and current results.

THE SELF-MANAGEMENT DRIVE

In his 1979 statement of goals, the CPS chairman said, "Let's meet the challenge of nine-waste elimination by increasing our self-management abilities." His goal was to clarify and develop the individual's role in the improvement process. He wanted to encourage planning at every level, even in indirect operations. Self-management activities at Canon promote individual skill development and responsibility in the context of a team effort for improvement. They are also designed to improve communication and communication skills.

Why was self-management considered so important? Every inquiry into the causes of work deficiencies or failures ultimately leads to people — someone who did not "do what's been decided." Only when everyone thoroughly understands his own work can this obstacle be overcome. Active communication and cooperation between supervisors and workers are required. Ideally, supervisors and workers discuss mutual goals and problem areas openly, workers develop effective methods and keep supervisors informed of problems and results, and supervisors provide the necessary support and resources.

More specifically, supervisors must:

- Make sure every worker understands the company's basic goals and methods.
- Monitor individuals' work and help manage the "gaps" between targets and actual performance.
- Make problem solving easier by discussing problems and countermeasures with workers.

Workers must:

- Set goals that are consistent with the supervisor's aims for the work center and draw up a personal goal achievement plan.
- Report results and problems to supervisor regularly.
- Aim for consistent self-management and execution of goal achievement plans.

Establishing a Basis for Self-Management

In Canon factories, section chiefs and foremen set up regular discussion sessions with their workers, either on an individual or a problem-area basis. These monthly or weekly meetings are used for group instruction, identifying objectives, problem solving, and sharing results. Open achievement records are also maintained for all workers to chart their progress in planning, productivity, merit eligibility based on skill examinations, number of work improvement proposals, and attendance rate.

Workers are expected to identify their own areas of concern and think through their own improvement measures and methods. They draw up individual goal achievement plans and put them into practice using self-management tools of their own design or the *Plan and Follow* (PF) sheets illustrated in Figure 5-2. These sheets can be used for recording results as well as for planning. Each day the worker records the results of work operations, problems encountered, and other pertinent information. This daily record is then used to prepare periodic reports for the supervisor, who responds with comments and suggestions.

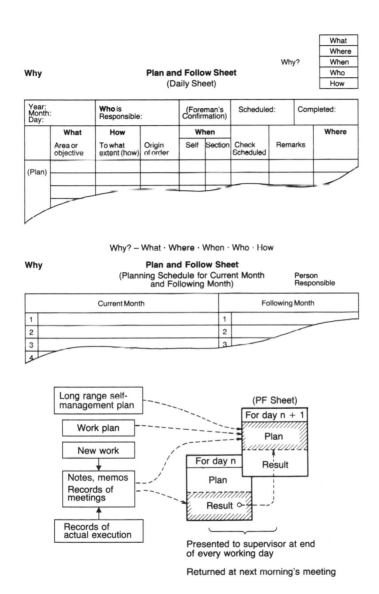

				What
				Where
			Why?	When
Why	Plan and Follow Sheet			Who
	(Daily Sheet)			How

Why? – What · Where · When · Who · How

- At end of working day, check your progress on the day's plan and enter in "Result" section.
- Look at day's results along with new work and previously scheduled plans, then enter new schedule in "Plan" section of sheet for following day. (Also use monthly sheet.)
- Report to supervisor at end of working day and present PF Sheets for both current and following days.

Figure 5-2 Self-Management Forms (PF Sheets)

The self-management program has had a number of important benefits:

- The program ultimately provides the knowledge and resources each person needs to solve any problem.
- Workers develop a consistent and rational work pattern: plan → execute → regulate → reflect.
- Workers who identify problem areas and solve them on their own have higher morale and a greater sense of involvement.
- Self-management enhances improvement activities: 300 people, each solving ten problems, produce 3000 success stories and greater overall improvement in the work centers.

Self-Management Improves Supervision

Self-management activities also improve supervisors' performance. For example, to help workers identify their own areas of concern, supervisors must observe them closely and attend to their needs regularly. Canon supervisors find that this activity has made them more skillful in handling their own problem areas. In addition, regular discussion of targeted areas opens up communication so that other kinds of problems can be solved smoothly. Finally, to provide effective guidance, supervisors must also check workers' progress frequently. Many new supervisors find this part of their work difficult, but once checking has become a comfortable habit, they become better leaders. Additional examples of self-management activities at Canon are described in Case Study 5.4.

CASE STUDIES

Case Study 5.1 — The 100-Item Improvement Plan at Factory U

"Problem consciousness" — the ability to pinpoint problems — is a popular phrase in Japan. But how does this consciousness develop? It is not enough simply to imagine an ideal condition and say, "I want my workplace to be like that!" Problem consciousness is born when we can define a current condition concretely (in measurable terms), calculate the difference between that condition and the ideal, and then identify the precise steps necessary to fill the gap.

As attention is focused on the gap between present conditions and a desired goal, the practical steps that must be taken become more apparent. As the goal is raised, the gap becomes greater, and specific tasks become more numerous and challenging.

Workers and managers sharpen their planning skills as they learn to identify and itemize the concrete steps required for any improvement. Developing this problem consciousness is the long-range goal behind the 100-Item Improvement Plan developed at Factory U.

At Factory U, all foremen are required to identify at least 100 tasks that must be done in their areas in order to achieve overall goals. And every section chief must identify and take responsibility for 200 more. The worksheet in Figure 5-3 illustrates problem-consciousness in concrete form. These worksheets are posted in the workplace so that new items can be added continuously. The vertical columns from left to right on the worksheet are explained below:

- *Waste:* The type of waste caused by the problem (defects, motion, work-in-process, equipment, human resources, etc.).
- *Problem:* The specific waste (the gap between present and ideal conditions), described in detail.
- *5M:* The 5M category (machinery, materials, manpower, method, or measurement) associated with the problem. [Editor's note: Like the concept of the nine wastes, the 5Ms are used universally to promote cause analysis in

100-Item Improvement Plan

◇ Study   ◆ Planning   ★ Implementation   Section:

Waste Category	Plan — Problem	5M	No.	Measure — Improvement	Person Responsible — Line	Person Responsible — Report	Schedule 82/1st half	Schedule 82/2nd half	Schedule 83/1st half	Schedule 83/2nd half	Estimated Results
Defects	G4 B surface defects	Machine	1	Turn over at lower position when polishing	Naka.	Ogawa	◇ Jan				Reduce from 1% → 0%
	CN-2 Stain 1366	Method	2	Protective membrane, heating method, finishing method	Naka.	Ogawa	◇ Jan				Reduce from 2% → 0%
	Stability of rough grinding process	Machine	3	Selection of pellets	Naka.	Tanaka	◇ Mar →				
		Method	4	Restudy amount of abrasion	Naka.	Ogawa	◇ Mar →				Reduce scratching 1.0% → 0.5%
	Crack defects	Machine	5	Change grindstone mesh on CG machine	Okama	Soto.	◇ Jan				
		Machine	6	Change cam	Okama	Soto.	◇ Feb →				(G4) 0.5% → 0%
	External defects overlooked	Man-power	7	Train and educate so all personnel can check this	Fuji.	Ouchi	◇ Jan, March				From 2% → 0.5% or less
	Need to increase precision of adjustment tool	Material	8	Change quality to SKS 31	Nagai	Okai	March				From 0.5% → 0.2% or less
	Precision lost rapidly	Method	9	Selection of rough polishing plate and polishing machine	Nagai	Tanaka	◇ Jan-Feb				
Motion	Heating to strip lens from plate takes time	Machine	1	Increase output of heating equip.	Nagai	Tanaka	◇ Jan				50 min. → 10 min./day
	One stripping worker must wait for other to finish	Method	2	Must position back and front sides of lens the same	Nagai		◇ Jan				
	Picking up material takes time	Machine	3	Lift lens with absorption pump	Nagai	Tanaka	◇ Feb				20 min. → 5 min./day
		Man-power	4	Use lost time on other operations	Nagai		◇ Jan				
	Too many checkers	Meas.; method	5	Make rounds at set intervals	Nagai	QA	◇ Jan-Mar				4 people → 3 people
			6	Improve method of measurement	Nagai	QA	◇ Feb				
	Too much machine trouble	Method	7	Diagnosis request to PM	Nagai	Inada	◇ Jan →				4 hr./month → 1 hr./month
		Machine	8	Change direction of nozzle	Nagai	Tanaka	◇ Feb				
	Want to shorten time for set up	Method	9	Use universal joint	Nagai		◇ Mar →				10 min./month → 5 min./month

Category	Problem	Type	No.	Improvement Item	Person	Schedule	Target
Work-in-process	Problems with handling modified parts	Method	1	Assign personnel to specialize in modifications	Ogawa	◇ Jan	Reduce 15,000 → 7,000
		Machine	2	Use specialized machines for heating and cooling	Ogawa Tanaka	◇ Mar ⟶	
			3	Select 2 workers to do modifications	Ogawa	★ Mar	
	Want to reduce storage of completed A-surface parts	Method	4	Improve double-coated protective membrane	Ogawa	◇ Jan	2,000 → 1,000
			5	Improve cooling rack	Ogawa	◆ Jan	
			6	Make easier to transport	Ogawa	◇ Jan	

Category	Problem	Type	No.	Improvement Item	Person	Schedule	Target
Equipment	Increase working ratio of lens centering machine	Method	1	Synchronize previous and subsequent processes	Fuji., Nakai	◇ Mar	Get by with 12 instead of 14 machines
				Plan load leveling			
	Floor of polishing machine gets dirty	Machine	2	Make oil receptacle easy to pull out	Ogawa Tanaka	◇ Feb ↑	

Category	Problem	Type	No.	Improvement Item	Person	Schedule	Target
Human resources	Lack of skill among people on second shift	Man-power	1	Use overlap time; carry out training	Ogawa	◇ Jan-Feb	
			2	Review training for workers in section	Ogawa Ouchi	◇ Jan-June	
	Information feedback needed	Man-power	3	Have second shift people write up individual observations (conditions, questions, items for inspection)	Ogawa Ouchi	◇ Jan ⟶ make this continuous	
	Dissatisfaction with work guidelines	Method	4	Draw up special work guidelines for second shift	Ogawa Ouchi	◇ Jan	

Figure 5-3   100-Item Improvement Plan/Example

improvement activities. "Problem consciousness" or "problem-finding" is not limited to Canon; a basic feature in the Japanese approach to improvement, it is practiced vigorously in many successful Japanese companies.]

- *No.:* Each improvement itemized by number.
- *Improvement:* Concrete action required to achieve the ideal or target condition (including consultation with related departments).
- *Person responsible:* Person who will implement the improvement.
- *Schedule:* Deadline for solving the problem or implementing the improvement.
- *Estimated results:* Calculating the projected results of improvement, which helps planners establish priorities for implementation. These figures can also be used later for comparison with actual results.

The following guidelines are adhered to in implementing improvements:

Calculation of effects. Whenever possible, quantify the expected results of improvement.

Liaison with staff departments. Unilateral action can lead to failure, so when on-the-spot solutions are not possible, approach staff departments for assistance.

When purely technical solutions are indicated, the assistance of production technology staff is required. Schedule regular meetings and make a clear division of tasks and responsibilities to ensure effective communication.

Progress checks. Visually indicate the status of improvement projects so everyone can keep track of overall progress. For example, at Factory U, improvement projects are marked in blue when begun and in red when completed.

Projects proposed in line-level improvement plans are routinely considered for inclusion in the annual goal achievement plans at Factory U. After factors such as priority, timing, and difficulty are considered, the basic concepts behind many of the projects are adopted and incorporated in the plans.

Case Study 5.2 — Three-Level Meetings and the N-Plan at Factory T

The N-Plan is a factory-wide design for overall improvement and goal achievement. The "N" in N-Plan stands for four English words used in slogans that promote its purpose. "To fulfill the PLAN and create a NICE (excellent) workplace, it is NECESSARY for NUMEROUS NEIGHBORS (fellow workers) to work together... Always think about what's truly NECESSARY and come up with NUMEROUS ideas for improvement."

In addition to small group activities emphasizing self-management, the N-Plan features regular planning and follow-through meetings at three levels — factory manager, division, and section chiefs. At these meetings all levels of management work together to establish consistent goals and objectives. (Figure 5-4)

Typically, the factory manager opens the meeting and presents his goals in the form of suggestions for the coming year. During his presentation, the manager supports his suggestions by discussing the factors to be considered: current overall state of the factory, the previous year's goals, and the requirements of the current three-year plan.

The participants discuss the goals together and offer their own views and concerns. When the manager's position is thoroughly understood and everyone's ideas have been worked in, a final statement is hammered out and approved by the manager. Later, at another group meeting, the goals of the division chief and section chiefs are completed in a similar fashion — each based on the plan above and each developed and approved through participation by all levels.

In these planning meetings, goals are always presented as reasoned suggestions or desires in order to leave the door open for differences of opinion and debate. As a result, the goals and objectives at each level are more thoroughly devel-

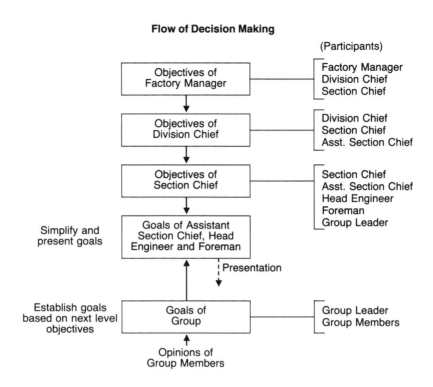

Figure 5-4 Setting Goals through Three-Level Meetings

oped, and managers at every level have a sense of shared purpose and commitment. The meetings have improved communication and cooperation in the factory in general, even between other departments. For this reason everyone's efforts are more easily coordinated toward achievement of the goals.

Case Study 5.3 — Inter-Level Management, Self-Management, and Three-a-Month Cards at Factory C

Inter-level management refers to the practice of monthly individual interviews between managers and staff or workers to discuss the current annual plan and important areas of concern. At these meetings, the manager or supervisor clarifies his or her general expectations and the specific tasks the worker is to fulfill. The worker also shares his or her own concerns, expectations, and goals for the month.

In Figure 5-5, "631" refers to monthly meetings between the factory manager, division and section chiefs to discuss goal achievement plans for the next six months, three months, and one month. Similarly, "321" and "21" refer to planning meetings between section chiefs and foremen, and between foremen and individual workers.

One-a-Month Agreement • One-a-Month Plan

Level	Interaction	Plan	Tools
Factory manager ←→ Section chief Division chief	One-a-month Agreement • Plan	631 management	631 plan 631 evaluation chart
Section chief ←→ Foreman		321 management	321 plan 321 evaluation chart
Foreman ←→ Workers		21 self-management	work planning schedule three-a-month cards

Figure 5-5 Inter-Level Management Process

The *21 self-management system* is a key feature of this system. It involves everyone in planning activity — from section chief to foreman to individual worker — and focuses their energies on plan achievement. The slogan "three-a-month" stands for the factory-wide commitment to increase productivity in all areas by 3 percent each month, with a target of 30 percent for the year. [Editor's note: The "three-a-month" or "3-Up" campaign has continued throughout Canon to the present. The level of monthly productivity increases is maintained through continuous improvements in all aspects of production — work methods, worker performance, operations, equipment, production, and product design.]

Three-a-Month Meeting I. At this preliminary meeting the section chief and foremen accomplish three important tasks:

- *Establish purpose:* They agree on specific details of section management for a 12-month period.

- *Give mutual advice:* They discuss the strengths and weaknesses of each block and decide what steps to take during that month.
- *Strive for unity of effort:* The group explores the possibility of new measures or policies to improve the section.

At the end of this meeting, a three-a-month planning proposal is drawn up to outline how the section as a whole should proceed to achieve its 3 percent productivity increase for the month.

The Three-a-Month Plan. Each foreman uses the three-a-month planning proposal to draw up an actual three-a-month plan for his unit. In this plan, he assigns target figures for groups and individuals in such areas as improving efficiency, reducing defect rate, work improvement proposal activity, Five S, and small group activities.

Three-a-Month Meeting II. In this meeting between foremen and workers, the Three-a-Month Plan for improvement is presented, discussed, and refined. Typically, the meeting occurs during the regular morning assemblies or over several meetings organized by problem area.

Three-a-Month Cards. These planning cards are the most important feature of the system. In one-on-one interviews, the foreman helps each worker understand the problem areas identified in the plan and tells him how he will be involved in the improvement activity. The worker writes this information down as his own plan on the three-a-month card. The card serves as a self-management schedule for each worker, so it will differ somewhat according to the individual's area of responsiblity, the work group, and the problems being addressed. However, all three-a-month cards are designed to include the following information (See Figure 5-6):

- monthly record of results
- daily results for current month
- end-of-month evaluation

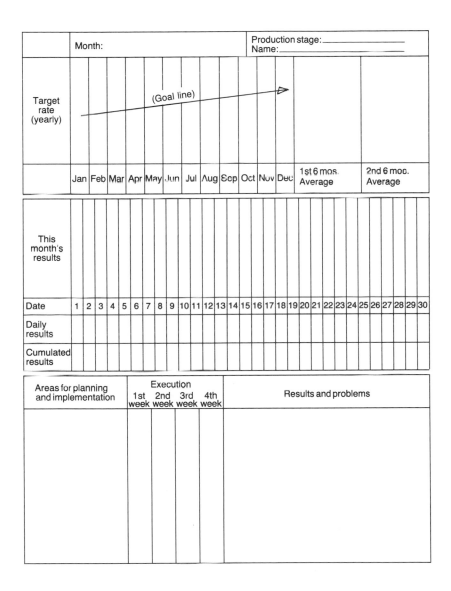

Figure 5-6 Three-a-Month Card Format

All entries are made by the individual workers, and their results are collected monthly and reviewed by the foreman and the section chief. The data are used in planning for the next month and as a basis for discussion in the next month's individual interviews.

Consistent use of this system month after month has produced many new improvement ideas at Factory C. Sharing detailed results through the cards has led to more consistent planning and has increased the waste elimination profit earned by the factory. Individuals find that their planning ability improves because the self-checks built into the cards help enforce the plan-do-see (PDS) habit. Day-to-day improvement efforts are now easier to coordinate, because everyone is used to staying in touch not only with group goals but with each other. Finally, the cards are a particularly effective tool for on-the-job training. New workers learn quickly what they can do, and the cards provide them with a daily guide and reminder.

Case Study 5.4 — Self-Management at Factory F: "My Areas for Improvement"

Vitality in the workplace is tied to the development of superior human resources. At Canon merit and individual aptitude have always been emphasized, along with the understanding that appropriate goals and a good system for self-management are the keys to optimal individual development. "My Areas for Improvement" is the name of a successful self-management program given high priority in the assembly sections at Factory F.

Background. In assembly operations, productivity is directly related to worker ability and efficiency. In this kind of workplace, high yearly increases can only be achieved through carefully planned and consistent management. Self-management is particularly important here: by making possible the fullest development of individual abilities, it directly benefits the work area and the company as a whole.

Developing worker skill — recording worker achievement. There are two important features of the My Areas for Improvement program:

- Workers' knowledge and skill levels are identified and developed through Hop-Step-Jump (Figure 5-7)

My Hop, Step, Jump M = manual available		
HOP		

General		Quality	
	• Ability to perform basic assembly operations M	31	• Understands following management charts: M
1	a) Tighten screws using a hand-held screwdriver	31	a) Histogram
2	b) Tighten screws using a pneumatic screwdriver	32	b) Parēto diagram
3	c) Tighten screws using an electric screwdriver	33	c) Special factors graph
4	d) Assemble parts using tweezers	34	d) P control graph
5	e) Perform gluing operations using an injector	35	e) $\overline{x}$ – R control chart
	• Ability to perform specialized operations M	36	Knows specifications for own work
6	a) Perform wiping operations with paper wrapped around tweezers	37	Knows function of parts used in own work
7	b) Do assembly operations using a G-ring tool M	38	Able to repair own work (changing screws, changing parts)
8	Understands materials used in own work	39	Contacts supervisors when any abnormality occurs
9	Carries out "My Areas for Improvement" program every month	40	Knows defect rate of own work in later stages of manufacture
10	Inspects tools and measuring devices used in own work		**Safety and Health**
11	Able to do own work according to work standards	41	Knows location of nearest stretchers and fire extinguishers
12	Knows names, part numbers, and unit cost of parts used in own work	42	Knows the distinctions among latent, work-stopping and non-working-stopping accidents
13	Able to take pictures with the camera being produced by own work	43	Knows how to move someone who is injured
14	Keeps personal locker in proper order		• Adheres to following soldering iron standards:
15	Keeps work area floor clear of parts and debris	44	a) Temperature
16	Knows CPS aims and the nine wastes	45	b) clean hands
17	Addresses others politely	46	c) Unplug when leaving work
18	Gives notice the day before an expected absence	47	d) The safe distance between soldering iron tip and eyes
19	Prompt and attentive for morning assembly	48	Knows health and safety slogans
20	Knows the goals set by the section chief	49	Does morning and 3:00 pm physical exercises
21	Knows the Ten Rules for assembly work	50	Knows the emergency exit routes
22	Knows Canon aims and Corporate principles	51	Knows color differences among solvent types 1,2,3
23	Begins work when the starting bell rings	52	Doesn't run in passageways or on stairs
24	Follows regulations regarding clothes, shoes, and nametag	53	• Knows the safety standards for "handwrap"
25	Maintains the 5 S's in own area		a) Countermeasures to prevent skipping
26	Able to change screwdriver bits		b) Distance from eyes
Productivity and Improvement			c) Cap use
27	Calculates own efficiency		**Education**
28	Performs own work 100%	54	Easy Improvement Methods
29	Submits suggestion memos	55	Earned 2nd Class certificate in soldering
30	Able to implement standards improvements for rubber gluing		
			HOP – 55 Items

Figure 5-7 My Hop, Step, Jump

- Workers' goals are tracked through My Areas for Improvement (Figure 5-8), their achievements through My Management Chart (Figure 5-9)

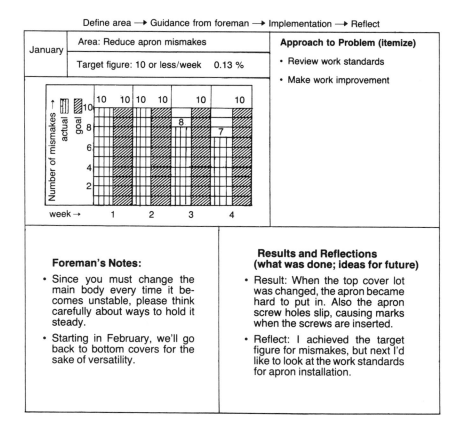

Figure 5-8 My Areas for Improvement

Hop-Step-Jump is a schedule of 130 items arranged in three levels of difficulty or importance, under the headings "Hop," "Step," and "Jump." For example, the ability to use a screwdriver and familiarity with the aims of CPS or the Five S principles are all "Hops." "Steps" might include learning proper use of grease and adhesives and understanding in-company specifications or the basics of quality control. To make a "Jump," a worker must learn the fundamentals of automation or industrial

My goals for the year:		Four Improvement Proposals per Month							
Productivity					Improvement Proposals				
Mo.	ST (RU)	Actual Measurement (RU)	Goal (Pf)	Pf (%)	Mo.	No. of proposals		No. of points	
						Goal	Actual	Goal	Actual
1	806	878	90	92	1	4	5	0.25	0
2	801	780	92	103	2	4	5	0.25	0.33
3	517	446	96	116	3	4	7	0.25	0.66
4	290	198	98	147	4	4	2	0.25	0.33
5	290	196	100	148	5	4	5	0.25	0.33
6	305	252	102	121	6	4	5	0.25	0.66
7	309	265	104	116	7	4	4	0.25	0.33
8	309	270	106	114	8	4	4	0.25	0.33
9					9	4			
10					10	4			
11					11	4			
12					12	4			

ST = Standard Time
RU = Ready Unit
Pf = Performance

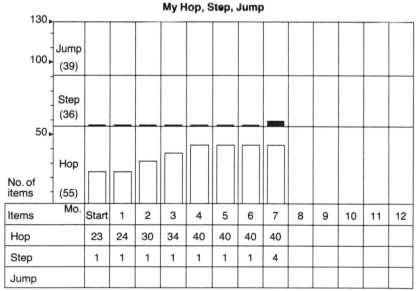

My Hop, Step, Jump

Items	Mo.	Start	1	2	3	4	5	6	7	8	9	10	11	12
Hop		23	24	30	34	40	40	40	40					
Step		1	1	1	1	1	1	1	4					
Jump														

Zero Defects (ZD) means doing work correctly from the
beginning and reducing mistakes to zero.

Figure 5-9 My Management Items

engineering or demonstrate leadership in improvement activity. The forms are designed to show the worker's progress at a glance.

On the monthly "My Areas for Improvement" form, a worker records the target figure for the improvement area she or he has chosen in consultation with the supervisor. There are many areas to choose from, e.g., reducing careless mistakes, increasing improvement proposals, or increasing efficiency. Each day the foreman takes a few moments to discuss with the worker his or her goals and achievements and any gaps between the two. Many new ideas and improvement proposals come out of these discussions.

Obviously, no foreman can have detailed daily discussions with 20-30 workers on a casual basis. To make the system work requires organization and discipline. Like the three-a-month cards described in Case Study 5.3, this system is a deceptively simple management technique. It requires an extra effort each day to make time available for the activity; in the long run, however, it makes everyone's job much easier. The system improves communication and worker skill; it also generates important management data. The section chief can stay in touch with issues being addressed on the line and take appropriate follow-up action on his own level.

Self-management activity is tied in with Canon's extensive company training programs. Training provides materials for self-study, as well as courses that range from one- to two-hour roundtable discussions to three-month intensive in-service projects. See Chapter 7 for a description of the training program philosophy and methodology.

Surprising Results of My Areas for Improvement. After the introduction of this program, some workers began coming to work two hours early to study for certification in work skills. Others stayed after work to discuss improvement ideas, even though their foremen explained that it was not required. No matter how often they were told "Don't come in so early" or "You'll be late getting home — call it a day!" they showed no sign of letting up. Their enthusiasm had led them to spend their own time discussing ways to raise QCD levels in their workplace.

Moral: If you want to set the workplace on fire, it's not enough to light a fire under the leaders. Only when leaders and workers are fired by the same enthusiasm is the workplace truly vitalized. ʼ

Keys to Fundamental Improvement

*Effective planning brings people together
to produce results.*

Even the best resources are useless unless they can be marshalled in a process that flows smoothly toward clear-cut objectives. Bringing people together to share information is the first step in the process; precise and continuous planning turns their combined know-how into concrete action.

Canon has systematized and harnessed the power of cooperation by giving people goals to strive for and reliable methods with which to regulate their progress. And, paradoxically, by emphasizing individual responsibility and self-development, the company has reaped the rewards of higher quality teamwork. The next chapter takes a more detailed look at the structured improvement activities Canon promotes to maintain high levels of participation and enthusiasm within its workforce.

6

Three Activities Aimed at
a Livelier Workplace

VITALIZING THE WORKPLACE THROUGH
THE "THREE SELFS"

The most popular and fundamental of the four Canon Principles emphasizes the threefold spirit of the self: self-motivation, self-respect, and self-reliance. This philosophy at Canon fosters an open, supportive environment along with a strong sense of duty, purpose, and responsibility.

The vitality of the workplace grows out of three morale-building activities based on the spirit of these "three selfs" — the *Five S campaign* (housekeeping for greater efficiency and reduction of waste), *small group activities*, and the *work improvement proposal system*. (Figure 6-1) Each of these activities lets workers experience the power of teamwork, creative endeavor, and success. Workers who have been buried in their work environment lose their passivity and become actively involved. They begin to think of ways to improve the work — it becomes interesting, enjoyable. And the vitality of the workplace increases rapidly.

This vitality is the most important force behind the success of the CPS effort, and it has fueled many creative improvements

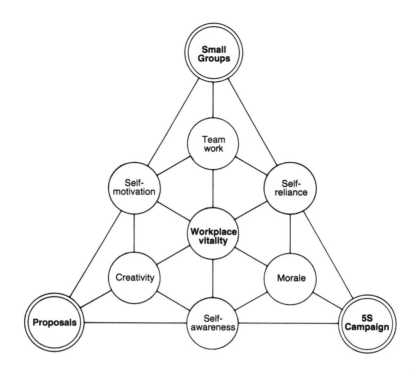

Figure 6-1 Three Activities that Vitalize the Workplace

at individual factories. Executives and management at Canon understand the importance of team projects and small group activities and take an active interest in them. Through their support, workplace vitalization activities have become a natural part of the daily routine at Canon.

THE FIVE S CAMPAIGN

It goes without saying that orderliness and organization are sensible as well as essential in the workplace. Improvement activities to promote these principles come under the general heading of Five S — five Japanese words that mean proper arrangement (*seiri*), orderliness (*seiton*), cleanliness (*seiketsu*), cleanup (*seiso*), and discipline (*shitsuke*). In some factories the English word "safety" adds a sixth S.

Some people complain that industrial housekeeping costs time and labor but produces no return. The function of Five S principles, however, is to reduce and ultimately prevent quality and productivity losses. They are essential to the elimination of waste in the work area and the factory as a whole.

Canon has benefited in two important ways through the Five S movement. First is the workers' change in consciousness — a readiness to follow the rules and "do what's been decided" that is now firmly entrenched throughout the workplace. For example, keeping parts and tools in their place was once a hard rule to enforce. It is never broken now; this reduces delays and wasted motion and helps visually control the workplace. The second benefit can be seen in the manufacturing process itself, in fewer accidents and equipment breakdowns, increased work efficiency, and lowered defect rates.

The Five S movement is promoted under the Canon slogan, "Use your legs and eyes; do regular checks and evaluations." For example, at Factories G and K, Five S Committees (chaired by the factory managers) conduct periodic inspections and photograph problem areas. The problems or suggestions for improvement are then reported on a Five S card with the photographs attached. The responsible work area must immediately develop a solution to the problem and submit a written plan to the Five S Committee.

Factory U has developed a self-grading Five S system. Each work area evaluates itself at set intervals (usually weekly), using a Five S checksheet. (Figure 6-2) These self-evaluations are reviewed by the foreman, section chief, and factory manager, then returned to the work area with comments or recommendations.

At Factories T and U, a work area that passes a certain level of inspection is designated a Five S Premier Work Center and awarded a Five S recognition plaque along with additional prizes. This work area then serves as a model for others in promoting the Five S movement. The model work areas must continue to pass regular inspections, however. They are expected to continue to look for ways to raise their level even higher.

Factory U

5 S Inspection Sheet	Evaluation Rank			Rank A: perfect score Rank B: 1-2 problems Rank C: 3 or more problems
	A	B	C	
Item				Comments
Proper Arrangement				
(Sort out unnecessary items)				
Are things posted on bulletin board uniformly?				
Have all unnecessary items been removed?				
Is it clear why unauthorized items are present?				
Are passageways and work areas clearly outlined?				
Are hoses and cord properly arranged?				
Good Order				
(A place for everything and everything in its place)				
Is everything kept in its own place?				
Are things put away after use?				
Are work areas uncluttered?				
Is everything fastened down that needs to be?				
Are shelves, tables, and cleaning implements orderly?				
Cleanliness				
(Prevent problems by keeping things clean.)				
Is clothing neat and clean?				
Are exhaust and ventilation adequate?				
Are work areas clean?				
Are machinery, equipment, fixtures, and drains kept clean?				
Are the white and green lines clean and unbroken?				
Cleanup				
(After-work maintenance and cleanup)				
Is the area free of trash and dust?				
Have all machines and equipment been cleaned?				
Has the floor been cleaned?				
Are cleanup responsibilities assigned?				
Are trash cans empty?				
Discipline				
(Maintaining good habits at Canon)				
Is everyone dressed according to regulations?				
Are smoking areas observed?				
Are private belongings put away?				
Does everyone refrain from eating and drinking in the workplace?				
Does everyone avoid private conversations during work time?				

Rank totals | | | |

Figure 6-2 Five S Inspection Sheet

SMALL GROUP ACTIVITIES

Small group activity programs flourish in every Canon factory, and everyone from line workers to plant managers participates in at least one group. (Figure 6-3) The names of group programs in each factory reflect their history or their primary focus. For example, the *ZD* groups in one plant got their start during the *Zero Defects* quality movement and kept the name. In another plant, *3G* activity refers to the three groups that must always work in close cooperation to ensure quality — the process in question and those on either side. Each group of three to eight or more workers has its own name or slogan to distinguish it from other groups, such as *Smile, Rainbow,* or *Sphinx.* (Figure 6-4)

Activity Level	Self-regulated Groups		Project-centered Groups	
Workers	• To increase profits for the corporation and skill levels for the individual. • To instill a feeling that the work is worth doing.	ZD 3G N Plan KS		
Foremen			• To increase productivity and reduce defect rates using IE and QC methods. • For mutual instruction and exchange of techniques.	Mutual analysis and counsel method • IE Plan 310 * • Scramble 10 * • 7-up *
Section Chiefs			• To improve factory manufacturing operations and management system.	Mutual analysis and counsel method • WD Plan *
Division Chiefs Factory Manager			• To plan innovation and vitalization in the factory as a whole.	Mutual analysis and counsel method • PI Plan *

* Each of these activities promotes company productivity improvement targets.

Figure 6-3 Focus of Small Group Activities by Level

Department	Name of Group (No. of Group Members)		Target	Target Value	Work Period	Result
Direct Labor						
Assembly Dept. 1	"Rainbow"	15	Shorten the time of an assembly task	150 RU	3 months	Achieved
Machining Dept.	"Perfect"	7	Reduce defective rate	5.2% ♦ 2.0%	6 months	Achieved (in 3 months)
Surface Treatment Department	"Smile"	11	Increase performance in models 500, 520	67% ♦ 90%	3 months	Achieved
Indirect Labor						
Supplier Management Dept.	"Suzume"	6	· Make new work standard · Improve present work standard	one item three items	3 months	Achieved
Purchasing Dept.	"Silk Road"	4	Reduce mistakes in issuing slips	less than 1%	3 months	Achieved (mistakes reduced to 0.35% for the monthly issued amount of 1800 slips)
Engineering Dept.	"Sphinx"	12	Prepare data base of total cost for different types of processing	Drilling Grinding Assembly etc.	3 months	Achieved

Figure 6-4 Report of Small Group Activity in Factory T (Targets Achieved)

Self-regulated small groups carry out each aspect of their activity independently, from the selection of themes to actual problem solving. (Figures 6-5 and 6-6) Quality, productivity, and cost reduction are the most frequently chosen themes. (Figure 6-7) The groups meet formally for two to four hours a month during regular working hours. The time they spend in meetings after work is also paid by the company. Group work is also supported by group activity offices in the factories and at company headquarters. These offices train groups and group leaders (see Figure 6-8), publish newsletters, and schedule open factory and company-wide meetings to present results and award prizes.

Groups achieving their targets receive a letter of commendation and a cash bonus from the factory manager. Groups with outstanding results are invited to present them at the All-Canon Small Group Activities Convention, held annually with affiliated companies. Since these conventions are important events, attended by both company executives and worker representatives, the honor of making an announcement is an excellent incentive to the groups.

Typically, a self-regulated small group consists of the workers within a single work center. Some groups, however, are made up of individuals from different work centers or sections who do the same kind of work. These groups have been particularly effective at Canon. Members bring their shared understanding to the problem, and their improvements can be implemented consistently in several work centers at the same time.

Case Study 6.1 — Miscellaneous Workers Improvement at Factory T

At Factory T, one worker in each section of the manufacturing division usually handles the miscellaneous jobs, involving paperwork and errands. These workers participated in their own section's small group activities, but felt out of place because no other member of the group did the same work. Furthermore, since the miscellaneous workers were typically

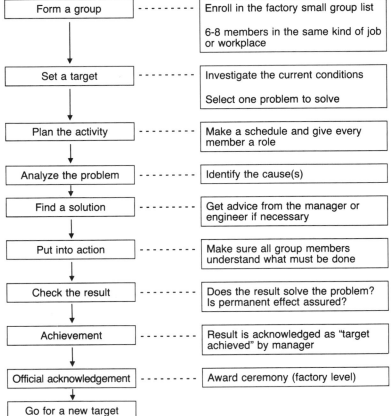

Figure 6-5 Group Activity Process

How to Set a Target

Identify current problems and select one to solve.

1. Is it a worthwhile problem to solve?

2. Can you expect support from your manager with the target you set?

3. Can the problem be solved through the effort of group members?

4. In tackling this problem, can you gain new experience and exercise your creativity?

Figure 6-6 Target-Setting Guidelines

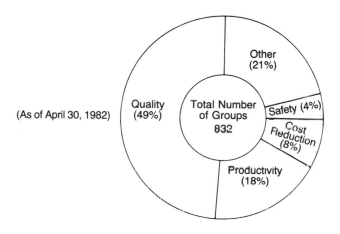

Figure 6-7 Theme for Small Group Activity

Courses for "3G" Program

Managers, Foremen

- Methods of Project Support
  (group discussion)
- Significance of Small Group Activity
  (group discussion)

Group Leaders

- Group Leader Training Course (4 hrs.)
  (use *Handbook of 3G Activity*)

- How to Hold Group Meetings (4 hrs.)

- Work-Improvement Methods (4 hrs.)

- Quality Control Methods
  (histogram, pareto diagram,
  graphs and charts)

- Significance of Group Activity

Workers

- Fundamental 3G (1 hr.)

- Work-Improvement Methods (4 hrs.)

Figure 6-8 Small Group Training at Factory T

women, they often felt isolated because other members of their group were all men. They did not lack motivation to tackle the problems in their own jobs, however. After consulting their supervisors, they formed their own six-member activity group.

The group's first step was to write their problems on cards, then classify and analyze them. They found that the biggest mutual problem area was paperwork management — copying, distributing, filing, etc. With their objective to "standardize paperwork and reexamine paper management," they set out to accomplish the following tasks:

- Determine what paperwork must be distributed and to whom.
- Standardize the format of documents used in all the sections.
- Create a unified system for classifying and managing paperwork.
- Establish the order in which paperwork is to be handled and clearly indicate where and for how long a document is to be filed.

This first project took the group eight months to complete — two months longer than they had planned. By the end of the project, however, the change in work procedures had produced significant improvements. (See Figure 6-9) Now 1.5 people could manage paperwork that formerly required 2.4 people; unnecessary duplication of copies was eliminated, resulting in a 20 percent reduction in filing binders; documents could be found quickly when they were needed.

Greater self-confidence was the activity's most important result, however. "We can do it, too," said the group members. "We designed management procedures for the entire manufacturing division." Their enthusiasm continued, even after one of them was moved to another position, because of the labor savings the group had achieved. Their next project was to develop a work manual.

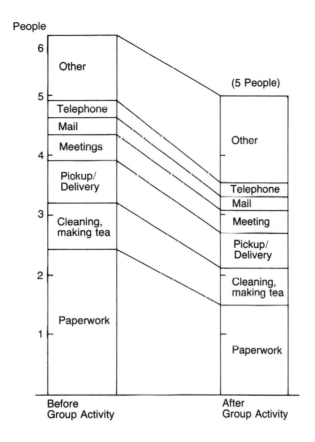

Figure 6-9 Results of "Miscellaneous" Workers' Group Activity

Mutual Analysis and Counsel Method

In addition to small groups at the worker level, Canon also encourages project-centered group activities at the foreman level and above. These groups have achieved excellent results that parallel the worker group activities in day-to-day operations.

In this type of group, a cross section of the plant's managers and supervisors reviews the problems in one department and investigates opportunities for improvement. On the basis of the advice and counsel received, the managers in that department then carry out the improvements.

Group members find they can:

- Look at the facts more objectively.
- Avoid falling into a rut, by looking at fresh approaches.
- Learn what works and what to avoid from the experience of other departments.
- Increase individual skills by sharing the very best ideas.
- Break down barriers to more effective communication and mutual evaluation in the future.

This method has proven so beneficial that it is now widely applied. For example, whenever company employees tour the workplace for inspection, they routinely write up their observations on a "mutual analysis and counsel sheet."

The following case illustrates how this method originated and how it works in practice.

Case Study 6.2 — Mutual Analysis and Counsel Method (IE Plan 310 at Factory F)

Ten years after the ZD movement began at Factory F, workers were still enthusiastically engaged in small group activities. Foremen and supervisors, however, were not involved in any organized improvement activity, and they saw this as a definite limitation. It occured to them that if all levels engaged in small group activity, the level of vitality in the factory would be much higher and more significant improvements could be achieved.

To meet this need, project-centered group activities were initiated at the foreman level and above. Even division chiefs and the factory manager became involved.

IE Plan 310 was a project to increase productivity in one work area by 10 percent within three months. A foreman from the section and two others arranged to put aside their usual work in order to devote themselves full time to the project for the first three weeks. Their activities fell into five broad stages:

Step 1. Understand present conditions — Do loss analysis in the target work area and determine where productivity can be increased.

Step 2. Analyze present conditions — List factors currently holding back productivity as problems to be solved and rough out a plan for improvement.

Step 3. Design the improvement plan — Using the previous analysis, brainstorm improvement ideas, evaluate them, and then draw up a detailed plan of action.

Step 4. Implementation — Communicate the plan to all departments involved and gain their cooperation, then implement the action plan.

Step 5. Confirmation — Do another loss analysis to quantify the results of all activities undertaken since the beginning of the project. (Figure 6-10)

After this activity began in May 1980, a new team was formed every month. Team learning and mutual instruction among supervisors and other managers substantially increased vitality among personnel at those levels. And this activity was eventually incorporated into Canon's IE in-service training for supervisors, as follow-up and reinforcement activity. (See Chapter 7)

WORK IMPROVEMENT PROPOSAL SYSTEM

The work improvement proposal system has a long history at Canon. It has been promoted in the company systematically since 1952. Its purpose is to increase interest and involvement in the work effort and promote harmony and cooperation among workers.

Areas of work improvement include:

- Production efficiency
- Cost reduction (expenses)
- Quality improvement
- Safety and hygiene
- Work center environment
- Individual work procedures
- Nine-waste elimination

1. **Target Workplace:** Unit Assembly, Assembly Section

    1. Workers: 12 people

    2. Workers move from place to place according to a work plan, performing independent unit assembly operations that differ from unit to unit.

    3. There are both automatic machine and hand assembly operations.

2. **Loss Analysis**

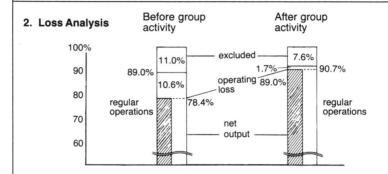

3. **Chief Measures Taken**

    1. **Shelf**
        1. Change construction
        2. Improve screw-tightening tool
        3. Discontinue pellet-check stage
        4. Improve work area

    2. **Spool gear**
        1. Shorten interval of automatic machine
        2. Rebuild 3St part feeder

    3. **142 ASA armature**
        1. Change layout
        2. Improve work area
        3. Shorten interval of air press

    4. **Overall**
        1. Take another look at work plan chart
        2. Thorough preparation before changeover.

4. **Graph of Changes in Productivity**

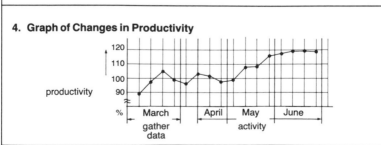

Figure 6-10 IE Plan 310 Activity—Sample Report

The Proposal Office was established to process suggestions from all personnel (except managers) — from development and design to manufacturing. When the CPS effort began in 1976, however, this office was relocated in the CPS Promotion Office so it could serve as one of the primary supports for the vitalization campaign.

Although CPS was aimed only at manufacturing, this move did not turn out to be problematic: after the transfer, proposal activity at Canon increased tenfold within the first two years and continued to rise, especially among line workers. (See Figure 6-11) [Editor's note: In 1982, the total number of proposals at Canon was 326,989, higher than the level projected in the

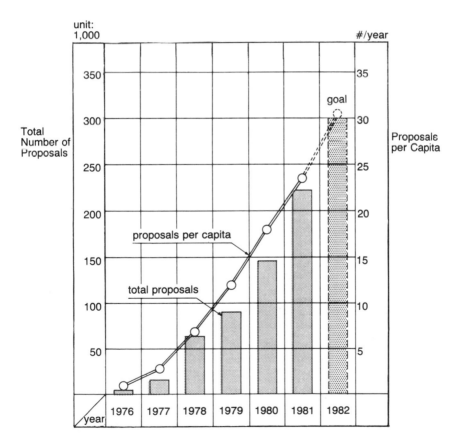

Figure 6-11 Increase In Work Improvement Proposals

graph. By 1985, the number had almost tripled, to 893,301. The per person rate had climbed to 70.2. The highest total submitted by one person was 2,600. In 1984 Canon was thirteenth in the top 20 Japanese companies with the highest total numbers of employee suggestions. Matsushita Electric Industries was in first place with over 6.5 million.]

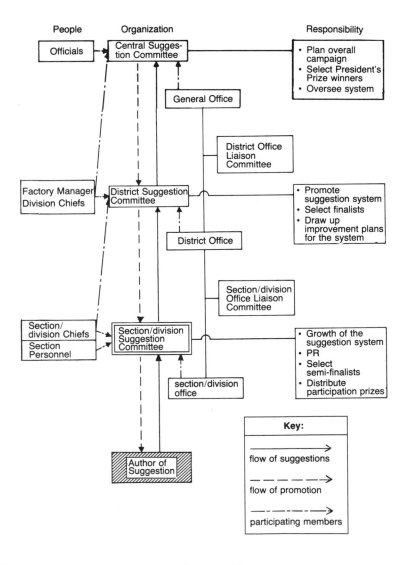

Figure 6-12 Work-Improvement Proposal System

One factor in this increase was the new proposal memo system introduced in 1978. Under this system, an improvement idea written in the form of a simple memo is accepted in lieu of a formal proposal. This simplified format helped many individual workers experience the excitement of seeing their own creativity produce change.

Another important factor was the role of the supervisors and managers in encouraging the development and implementation of suggestions. Managers at Canon do not participate in the proposal system — their contribution is to give workers the instruction and guidance they need in order to come up with increasingly effective and creative ideas. The role of the manager in work improvement is to:

1. Show a positive attitude
2. Give hints to workers
3. Listen to workers' problems
4. Make improvement targets clear
5. Plan competitions and games
6. Put proposals into practice quickly
7. Give public recognition

What is Not a Proposal:

1. A wish or a demand.
   Example: I want a raise.

2. An abstract idea.
   Example: The company needs a new product.

3. A complaint.
   Example: The food at the cafeteria is bad.

4. Personal criticism.
   Example: I don't like A. He should be transferred.

5. Something that benefits only certain people.
   Example: We want an air-conditioner for our room.

In each factory proposals are judged by the section committee, and the grades of honorable mention through E Prize are assigned immediately. (Figure 6-12) All suggestions are recognized: even those that don't receive an honorable mention (acceptance award) are given a 50 yen participation certificate. (See Figure 6-13)

Proposal Rank	Prize
Special President's Award	Trip Abroad
A	¥ 50,000 ($250)
B	¥ 20,000 ($100)
C	¥ 10,000 ($50)
D	¥ 5,000 ($25)
E	¥ 3,000 ($15)
Good	¥ 1,500 ($7.50)
Fair	¥ 500 ($2.50)
Acceptance Award	¥ 150 ($.75)

Figure 6-13 Proposal Ranks and Prizes

Proposals considered eligible for higher prizes (D through A) are formally re-submitted and judged by the district committee.

The implementation rate for suggestions is high — over 90 percent. In fact, most suggestions are not even submitted as proposals until after they have been implemented successfully. After all, the rationale for the system is not simply to gain recognition, but to make successful improvement ideas work for the entire company. [Editor's note: By 1985, the total savings to Canon from proposals rated above the level of "Fair" was over $202 million. The operating expense of the system was only $2.2 million, 96 percent of which was prize money.]

Here is an example of a proposal to promote improvement:

Improvement Time — every day from 11:30 to 12:00. The foreman can use this time only for activity related to improving the workplace. He may not attend meetings or answer the phone, and his colleagues should also avoid this period in arranging meetings. The foreman must use this time to think about problems in an organized way and draw up plans that will lead to truly effective improvements.

Once a year around ten of the year's A and B prize proposals are chosen to compete for the President's Prize for Work Improvement Proposals. The two with the highest ratings receive the Special President's Prize, and the authors are sent on an overseas study tour. (See Figure 6-14)

One of the most notable features of the Canon suggestion system is the prize for lifetime accumulated points. Anyone submitting a proposal awarded an honorable mention or above receives from .33 to 5.0 points, depending on the grade received. These points are cumulated, and every year the 20 people with the highest totals receive the President's Prize for Accumulated Points, along with a gold medal and 300,000 yen ($1,500). The 30 people earning the highest points that particular year receive the President's Prize for Yearly Points, which carries an award of 100,000 yen ($750) and a silver medal. These annual prizes are a great incentive to authors of suggestions and a great source of pride to those who win.

THE AWARDS SYSTEM

One of the ideals of the CPS is to ensure that "evaluations are open; opportunity to win is equal." For this reason, personnel at every level from section chief down may compete for substantial prizes and announce their achievements regularly in a special forum.

The highest CPS award is the Premier Work Center Prize, given to sections practicing outstanding day-to-day management and achieving superior results in waste elimination. In each factory the CPS chairman personally inspects the sections

Prize	Awarded to:	Achievement	Amount	Additional prize	No. Awards	Announced at:
Premier workplace	Section	• All-Canon model workplace • 30% improvement rate 3 years in a row	$1,000	Overseas study tour for section chief Eagle shield (gold)	1-2	CPS convention
Runner-up premier workplace	Section	• Factory model workplace • 30 % improvement rate 3 years in a row	$500	Eagle shield (silver)	10 +	
Excellence in waste elimination	Block, team	Excellent results in waste elimination	$250		50	
Prize for increasing CPS results	Foreman, asst. section chiefs, specialists	Successful application of waste elimination methods		Overseas study tour	3	CPS waste elimination convention CPS convention
Excellence in small group activities	Group	Excellent results through small group activities	$250		2	Canon small group activity convention
President's prize for accumulated points	Top 20 people	Highest total points for work improvement proposals	$1,500	Gold medal	20	
President's prize for yearly points	Top 30 people	Highest annual points for work improvement proposals	$500	Silver medal	30	
President's prize	Individual, group	Best of the year's proposals, B prize and up	$500		10 +	
Special president's prize	Individual, group	Top two president's prize winners		Overseas study tour for representative	2	CPS convention
Gold quality assurance prize (for cooperating factories)	Cooperating factory	Most outstanding quality record	$1,000	Eagle shield (gold)	2	Cooperating factories management training meeting
Silver quality assurance prize (for cooperating factories)	Cooperating factory	Outstanding quality record	$500	Eagle shield (silver)	3	

Figure 6-14 Awards and Honors

recommended by the factory managers and interviews the section chiefs before making his decision. The winning section receives a money prize and a golden eagle shield (Photo 6-1), and the section chief is sent on an overseas study tour.

Photo 6-1 Golden Eagle Shield

The Prize for Excellence in Nine-Waste Elimination is given to work areas or teams achieving outstanding results in each of the nine types of waste. Other awards include the Prize for Increasing CPS Results, awarded to foremen and deputy section chiefs, and the previously mentioned Prize for Excellence in Small Group Activities. The President's Prize for Accumulated Points, the President's Prize for Yearly Points, the Special President's Prize, and the President's Prize are all related to work improvement proposals. Finally, individual factories have a variety of award programs of their own.

At Canon the award system is truly open. People who work hard to produce excellent results can expect to be highly praised and rewarded. Evaluation and selection are held to the strictest standards, however — in some years, no one is eligible for certain prizes. On the other hand, those who make the effort can and do win prizes over and over again.

ANNOUNCEMENT CONVENTIONS

Announcement conventions (Photo 6-2) serve three basic functions: to select outstanding personnel or work areas, announce the results of activities, and provide a forum for mutual instruction. (See Figure 6-15)

Photo 6-2 Announcement Convention

The CPS Waste Elimination Results Convention and the All-Canon Small Group Activity Convention fall into the first category. Every year local announcement meetings are held at every factory to decide who will represent the factory at the company-wide conventions. At the convention these individuals are given the honor of making a 20-minute presentation on their improvement activities and vying for the Prize for Increasing CPS Results and the Prize for Excellence in Small Group Activities. The winners are then selected and announced by the CPS chairman and the Study Group Committee.

The CPS Convention is the largest forum for the second category, announcement of improvement results. Held with

Announcement Convention	Individuals Making Announcements	
CPS Waste Elimination Convention	Foremen, asst. section chiefs, and staff	People contributing to successful workplace management through new waste elimination methods.
Canon Small Group Activity Convention	Small-group activity group leaders: (1 person from each factory)	Groups achieving outstanding results in small group activity
CPS Convention	Premier workplace prize: Section chief of winning section	Section chiefs
	Prize for increasing CPS results: 3 prize winners	Foreman, deputy section chiefs, staff
	Prize for excellence in small group activities: Leader of winning group	General workers
	Special President's prize for work improvement proposals: 1 prize winner	General workers

Figure 6-15 Announcement Conventions

considerable fanfare every spring, this convention is attended by the president and other top company officials and sums up the results of the CPS activities for the previous year. All the section chiefs, foremen, ordinary workers, and staff who received awards during that year present their activities and achievements and are honored by the company as a whole.

Canon's announcement conventions bring together and reinforce all the vitalizing activities described in this chapter. "Conventions where everyone can learn together" is an apt slogan to describe them, because they do more than serve as forums for reporting work activities and achievements. Those selected to speak can learn from having to reflect on their activities and summarize them for presentation. Those who listen have the opportunity to learn new management techniques and gain insight from the experience of other factories, sections, and individuals. And, more important, everyone learns *directly,* as part of a vivid personal experience. Finally, Canon benefits as a whole — in the higher levels of achievement for Canon factories and work areas, and for the individuals who work in them. For this reason, representatives come to listen not only from every Canon factory, but also from affiliated and overseas factories.

Keys to Fundamental Improvement

Continuous improvement is the result of continuous involvement.

As long as improvement activity is viewed as "extra" effort, improvement results will only be "extra-ordinary," that is, periodic and inconsistent. It is not enough simply to maintain high levels; the goal must be to exceed them — always.

Canon established this "habit of improvement" by making improvement activity an inseparable part of everyone's daily work.

Opportunities for improvement are provided in a structured way through Five S and small group activities as well as company-wide waste elimination and productivity improvement drives. Incentives for individual and group efforts are provided through the work improvement proposal award and announcement systems. This wide range of activities satisfies individual as well as group concerns and helps achieve company targets at the same time.

Individual involvement at Canon is not maintained entirely by the appeal of group goals or satisfaction of group activity, however. At every level, opportunities exist for self-development and advancement through formal and informal skills training. Chapter 7 identifies the basic model for Canon training programs by taking a detailed look at CPS training.

7

A Training Program to Support the Waste Elimination Effort

THE ROLE OF TRAINING IN PRODUCTION

Canon is served by a number of independent training departments. (Figure 7-1) This chapter, however, will focus on CPS training programs. CPS has three important objectives in training:

- To promote the successful implementation of the production system;
- To support annual goal achievement by teaching a team approach and reliable methods; and
- To raise the overall level of human resources by fostering individual responsibility, self-management, and self-improvement.

Training to Promote CPS

Keeping everyone informed about CPS operations and activities is not easy. Courses in CPS basic operations must be taught continuously. Improvement techniques developed and applied successfully in one work center must be made available to others. Work centers striving for premier work center status

Training Center (Skill Development Section)	Planning and development of company-wide human resources development system
Development Division	Planning office for technology training: Human resources development for research and development division
Production Department	CPS training office: Human resources development – production division
Overseas Marketing Division	Overseas operations management section: Training plans for overseas marketing personnel

Figure 7-1 Training Departments

need information about the staff support that is available. In addition, workers understand their own work better when they can learn more about Canon products and the functions of other company departments.

In the past, this kind of education and information sharing happened informally — through personal contact and word of mouth, in much the same way family members transmit information to each other. The informal method made an important contribution in its own time; however, the number of workers in the production department in Japan alone has grown to over 10,000. When CPS was introduced, ways had to be found to give a sense of direction and self-regulation to those involved in CPS activities. More efficient and standardized methods of communication were gradually developed, among them the training materials and publications shown in Figure 7-2.

Training to Support Annual Goal Achievement

Three of the nine wastes — work-in-process, defects, and motion — must be confronted in the workplace. Furthermore, they cannot be dealt with adequately by individuals — not even by individual work centers. Only company-wide teamwork brings results. That essential team approach is emphasized in every Canon training program, along with methods and practical activities to promote goal achievement.

CPS promotion office (company headquarters)	**CPS textbooks** **CPS slides**	Introduction to CPS activities
CPS training office	**CPS News**	Monthly status report on Company-wide CPS activities
Company headquarters in cooperation with factories	**CPS Handbook**	Basic CPS study guide (revised annually)
Individual factories	**CPS Bulletin**	Bi-monthly status report on CPS activities in the factory
Larger sections	**CPS Bulletin**	Annual statement of section goals

Figure 7-2 CPS Training Publications

Photo 7-1 CPS Training Publications

For example, as indicated earlier, a 3 percent monthly increase in productivity is expected from every work center in each factory as part of the CPS effort. The training methods described later in the chapter illustrate the important supporting role played by training in achieving this goal.

Figure 7-3 CPS Waste Cat

Training to Raise the Level of Human Resources

In any operation focusing on improvement and teamwork, ready-made plans devised by a "super-manager" or handed down from on high are not desirable. Although it may appear less efficient, Canon policies and practices are established collectively, with contributions expected from everyone. This approach helps maintain the level of vitality (individual involvement and commitment) needed to make rapid improvement. This means, however, that Canon has a great deal invested in the *future* skill and abilities of its employees. The more opportunities for growth are provided, the greater the quality of individuals' contribution can be. Thus, training is seen as an opportunity for self-development — an independent source of vitalization within the workplace.

DEVELOPING A CURRICULUM FOR THE PRODUCTION DEPARTMENT

During the ten years preceding the Premier Company Plan, Canon grew to two-and-a-half times its former size — too large and complex to just "let people grow naturally." And, although one of the best ways to train people is to move them around, there were also limits to what could be accomplished through training by rotation. Some departments were putting more energy into training than others; and in work centers that were strong in a particular area, training progressed more rapidly than in others. To eliminate these inequalities, Canon decided to initiate a company-wide leveling process. The new training program would incorporate the following principles:

- Develop basic training that is structured, practical, and reliable.
- Combine training organized by level with training organized by function.
- Make training part of the overall plan for human resources development.
- Always provide follow-up systems so training can be linked to work results.
- Focus training to support QCD control.
- Establish unified objectives for training at all levels.
- Use the "story" training method, which combines classroom and on-the-job training (see below).
- Develop company teaching staff by using outstanding company personnel as teachers.
- Establish standards for programs so differences in training opportunities do not develop betweeen factories or divisions.

To put together a comprehensive training program, representatives from every factory met together to examine the task from all sides. In addition to the inequalities cited above, further discrepancies were found between departments and work centers in the use of off-the-job (classroom) training and in human resources development planning in general. Training curriculums were standardized according to the following principles:

1. To ensure a human resources development program aimed at all personnel, clarify requirements and eliminate inequalities in training opportunities.
2. Since training is handled independently at each factory according to its needs, draw up guidelines to set consistent levels for training.
3. Wherever possible, incorporate the strengths of existing programs into new standards (particularly in on-the-job training).
4. Make sure everyone understands that human resources development is a continuous process.

The standard curriculum changes somewhat every year. A recent curriculum for management training and control in-service training is shown in Figure 7-4.

PROGRESSIVE OR "STORY" TRAINING

Training in the past did not always go beyond instruction in theory and methods. Practical application was left up to individuals and the work center. And, as mentioned above, the quality of on-the-job training varied greatly between divisions and work centers. As a result, many supervisors did not have a standard procedure to follow to implement improvements efficiently.

The CPS "story" training method (as in the stories of a building) was first used to overcome this problem. In the story method, on-the-job training becomes a part of the instructional design. Theory, methods, and practical application are taught successively as steps in a process of problem investigation and planning for improvement. The following example illustrates how this method was used to develop a successful IE in-service training for supervisors.

Case Study — The "Story Method" in IE In-Service Training

On the production line, waste caused by defects, waste in motion, and waste caused by work-in-process are the biggest concerns. Results achieved in these areas are due to the combined efforts of line and staff personnel. Training has supported these achievements, however, by standardizing the best improvement methods.

For example, in 1971 Canon introduced PAC (Performance Analysis and Control). PAC promoted higher productivity through control activities that involved repeated observation and guidance by foremen. As a result of the foremen's efforts, waste elimination attributable to PAC accounted for about 30 percent of the results in the first year of the waste elimination drive; in the following year it accounted for almost 40 percent.

For the foremen's collective effectiveness to produce continuing results, however, their methods had to be standardized at some point. The very best had to become more than the upper limit of the performance range; it had to serve as a model for everyone's practice.

With this objective in mind, the existing pattern of training for foremen was replaced, in 1977, with *IE In-Service Training for Supervisors,* a program developed with the help of the Japan Management Association.

The IE in-service training is an excellent example of the story training method used at Canon. Instead of "spot" training — theory and practice taught separately — concepts, methodologies, and practice are taught as steps in the solution of *real problems.* Following these steps, training begins with problem identification and investigation and then proceeds to actual planning, implementation, and analysis of results. The participants learn one step at a time, ascending the stairs from one "story" to the next. (Figure 7-5)

Foremen first practice on paper, learning how to identify and measure actual conditions in the work centers, the proper sequence to follow in making improvements, and the best

Production Department Standard Curriculum

A. Management Training

Level/Area	Work Responsibilities	Shape Up	Goal Management
New Employees	• Orientation • Basic Training (college grad.) ◇ • Basic Training (HS grad.) *		
General	• 2nd Year Basic ◇ • Intermediate Technician ◇ (with Core) • Workmanship Training*	Shape Up (with Core)	
Higher Grades (Core workers)	• Core Basic ◇ • Intermediate Technician ◇ • Core Skilled ◇ • Skills Inventory * (with foremen)	Core Shape Up	Goal Management (with foremen)
Foremen Specialists	• Foreman Basic • Management Training ◇ • Foreman's Planning • Skills Inventory*	Foreman Shape Up	Goal Management
Higher Grades	• Management Trainee Basic ◇		
Section Chiefs	• Self-Assessment • Pre-Manager Course *	Section Chief Shape Up	Goal Management Section Chiefs
Higher Grades	• Management Trainee		
Division Chiefs	• Executive Basic	Division Chief Shape Up	Goal Management Division Chiefs

Sponsoring Departments: ◇ = Training Center, * = Individual factories, all others = CPS Training Office

Shape Up: program fostering innovative, positive thinking

B. In-Service Control Training

Level/Area	Quality Assurance	Cost Assurance	Production Assurance	Common Courses	Computer
New Employees				Basic CPS*	
General	Introductory QC	VE Basic* (through foremen level)	Easy Improvements* IE Basic* (with Core)	• Management Game (all levels - Intro and Specialist) • Leading Groups • Group Creativity (all levels - brainstorming)	Office Automation (In-service)
Higher Grades (Core)	Basic QC Methods* (with foremen)	CE Basic (through foremen level)	IE Planning* (Scramble 10) (with foremen) IE Basic*	Training-within-Industry (with foremen) • Work Relations • Safety • Work Methods • Work Instruction Introductory Accounting (with foremen)	
Foremen Specialists	Supervisors QC Basic QC Methods*		IE In-Service Supervisors IE Planning*	• Safety, Health Laws • Meeting with Chairman • Fostering Creativity • Handy PERT • Interactive Modelling (IM) • Training-within-Industry	
Section Chiefs	QC Section Chiefs	VE Section Chiefs Cost Management	IE Section Chiefs	• Section Chief Basic • Kepner-Tregoe • Promoting IM • Duty Management/Labor Relations ◇	
Division Chiefs	QC Division Chiefs	VE Division Chiefs	Production Assurance Division Chiefs	Management Simulation	Office Automation (Promotion)

Figure 7-4   Example of Training Curriculum

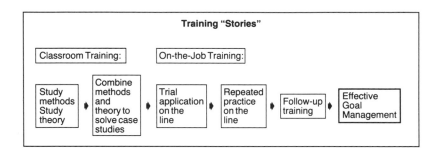

Figure 7-5 "Story" Training: Step-by-Step Theory and Practice

methods to use. (Figure 7-6) Then they apply what they have learned in actual practice on the line.

At this stage, the foremen investigate the possibilities for increasing productivity in their own work centers; then they draw up improvement plans based on this analysis. As a final step, the plans for increased productivity are presented at joint announcement meetings in the work centers.

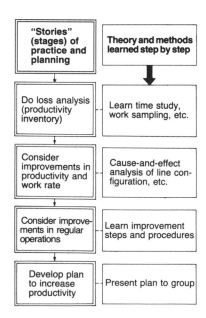

Figure 7-6 IE In-Service Training for Supervisors

Organization of the Training Program

During the 32-day program, the foremen devote themselves full-time to the training. In their absence, group leaders
and other key line workers fill in to keep up morale on the line.
The foremen are not completely out of touch with their work,
however, since they spend some of their time on the line each
week during the training practice periods. (See Figure 7-7)

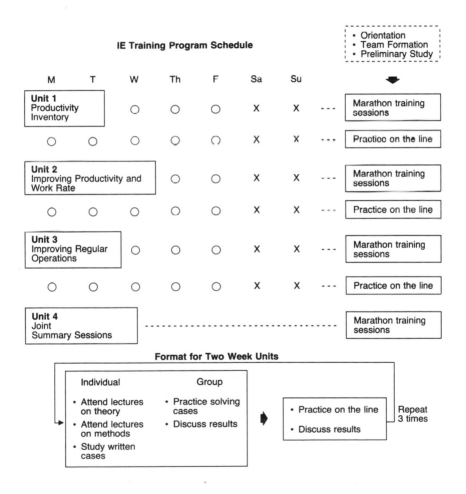

Figure 7-7 IE Training Program Schedule and Unit Format

Follow-up Training

It is only when the techniques learned in training can be practiced consistently and easily that individuals are able to increase their ability and achieve higher results from their improvement efforts. For this reason follow-up programs are strongly encouraged and supported by CPS, especially in supervisor training. The IE training process of investigation, planning, and implementation for productivity improvement is woven into the daily activities of management and supervision and reinforced in special follow-up projects.

The names of these follow-up programs differ from factory to factory ("IE Plan 310," "Scramble 10," etc.), but the methods and goals are essentially the same. Participants are given written "appointments" to devote themselves exclusively to special improvement projects that run for three weeks of investigation and planning.

Training overview for section chiefs. A work center that is not self-regulating or orderly can have a serious impact on other work centers. For this reason, the supervisor's role is vital — his day-to-day management, instruction and support of workers, and cooperation with staff departments help the work center fulfill its role in a continuously flowing process. Support and assistance from upper management are also important, however. To ensure effective follow-up and support, section chiefs are required to attend a three-day (and two-night) overview of the training program attended by their foremen.

Results of the training program. Figure 7-8 shows an example of the results achieved in the IE in-service training program, selected from among those presented at the joint announcement meeting that concludes the program. As the example illustrates, the foreman has learned to express in figures both the actual overall workplace productivity rate and the potential for increasing productivity. Since it is now possible for him to calculate when and how much improvement can be anticipated, his planning to meet annual productivity goals is likely to be more accurate.

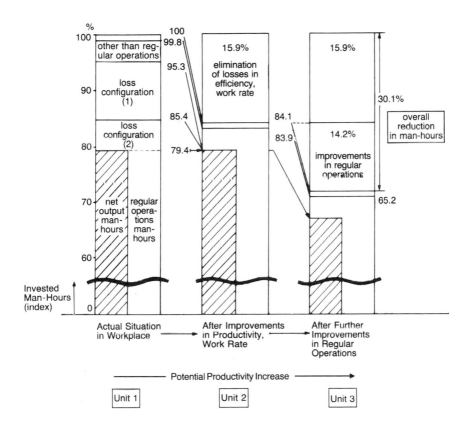

Figure 7-8 Projected Results of Improvements Due to Loss Analysis

Over 600 foremen and section chiefs were trained between 1977 (when the program began) and 1982. The methods and tools provided in the program gave supervisors, managers, and staff personnel a common language to use in planning productivity improvement. The greater efficiency this produced in both work and communication has paid off in steady productivity increases that continue today.

Keys to Fundamental Improvement

Give people training they can use — to do work that is personally challenging and important to the company.

CPS training at Canon follows the model illustrated in the IE in-service training program. (Another successful program based on this model is the VEWSS value engineering workshop seminar program described in Chapter 10.) The best methods are standardized and presented in a learning format that blends the benefits of classroom instruction and on-the-job training. The model promotes continuous improvement and successful goal achievement by giving people tools and methods they can use right away. By raising and leveling the quality of training available to Canon personnel, the company has achieved higher, more consistent results in quality and productivity improvement.

Chapters 2 through 7 have outlined the improvement software — the strategies and management systems used by CPS to support Canon's commitment to achieve world-class status under its Premier Company Plan. Chapter 8 also describes some of the improvement hardware developed and implemented under CPS in the HIT just-in-time production system.

8

Assuring Delivery Time Through the HIT Production System

Canon was originally a camera company, but when it began to produce business machines, production conditions and requirements changed dramatically. Business machines are much larger than cameras, and varying electrical requirements made it necessary to produce a variety of models.

At the same time, market demands had become more diverse since the oil crisis. It was clear that Canon needed a manufacturing system that could adapt quickly to changes in the market and produce quickly just the quantities Canon was able to sell. (Figure 8-1)

Internal and external factors such as these triggered development of the HIT production system — a continuous flow system flexible enough to respond quickly to change, and one that could significantly reduce production time and ultimately product cost.

THE HIT PRODUCTION SYSTEM — OVERVIEW

The HIT system is Canon's Just-in-Time parts supply. The name "HIT" comes from the first letters of three Japanese phrases that mean, "Make what is needed, when needed, in just

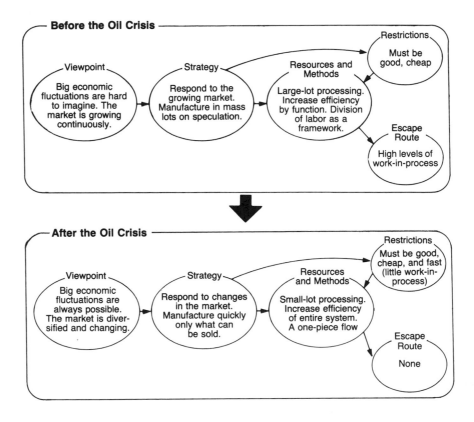

Figure 8-1 Changing Production Strategies

the amount needed."* As shown in Figure 8-2, HIT production consists of a continuous flow system combined with Canon's own approach to process control and work management.

In an ideal continuous flow system, work flows without accumulation or work-in-process. To achieve this, Canon implemented improvements in four major areas: systematization, small lot size, load leveling, and pull-by-subsequent-process.

In process control, Canon developed two methods to make sure the right products were made when needed and in the amounts needed.

---

* H — *Hitsuyo na mono o* (what is needed)
  I — *Iru toki, iru dake* (when needed, in the amount needed)
  T — *Tsukuru* (make)

HIT cards (Canon's version of kanban) were introduced in mass production factories to maintain the pull system and control the level of work-in-process. (See pp. 132-139) In factories with low-volume, mixed model production, the *signal system* was introduced — a computer-supported information system designed to prevent shortages and maintain daily schedules. (See pp. 147-153)

Finally, in the work management area, Canon undertook a comprehensive waste elimination campaign. Productivity rose through work improvements and improvements in regular operations. However, Canon also increased productivity by using standard work guidelines, PAC, and visual control systems to help workers achieve higher performance levels.

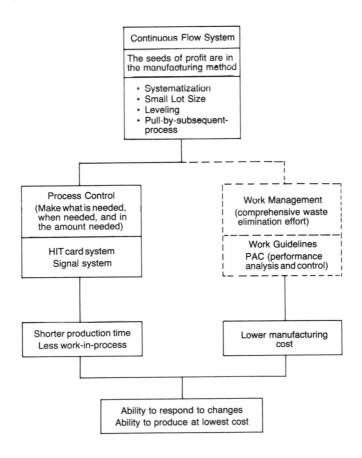

Figure 8-2 Basic Principles of the HIT Production System

Organizing for Improvement

To realize CPS goals for improving the production system, each level of the organization was given responsibility for different areas. (Figure 8-3) Middle managers promoted the implementation of the HIT system and directed the creation of synchronized, stockless operations and processes. Foremen and supervisors were asked to promote productivity improvement in the work centers through PAC. They researched efficient methods, established work guidelines, and provided instruction and guidance to workers. Workers were expected to follow the work guidelines and promote workmanship and output within the system.

Canon's guiding principle in promoting CPS and the HIT system was a simple one: *give people work worth doing and the tools they need to make improvements — then watch how self-*

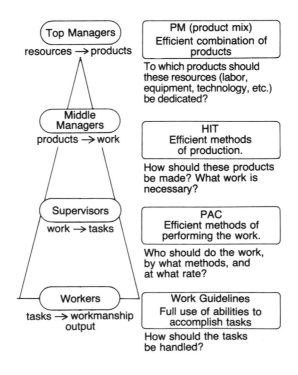

Figure 8-3 Making the HIT Production System Work

confidence and competence increase along with concrete achievements. (See Figure 8-4) The improvements described in this chapter illustrate not only the success of the CPS plan but also the overall growth of people working toward objectives that were important to them personally and as a team.

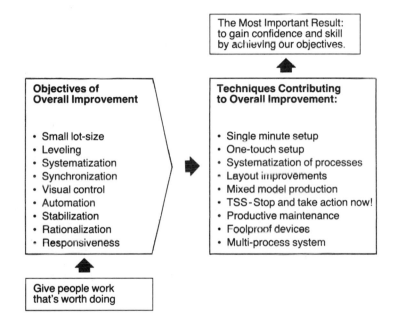

Figure 8-4 Overall Objectives of the HIT Production System

Features of the Continuous Flow System

As mentioned above, four kinds of improvements must be implemented to promote a continuous flow system:

1. *Systematization:* Arranging processes so that a part can be processed and assembled in a single sequence with as little delay or distance between processes as possible. Work center organization corresponds to the flow of the process.

2. *Small lot size:* Shortening setup time to make frequent changeovers feasible and small lot production economical.

3. *Load leveling:* Averaging production volume and variety to reduce fluctuations in process load (smoothing production).

4. *Pull-by-subsequent-process:* Drawing parts from the previous process only when needed and only in the amount needed.

Systematization and small lot size make it possible to achieve a flow without accumulation, i.e., no waiting time for parts between processes or during processing. They also facilitate synchronization. That is, the timing and volume of processing in the previous process coincide with those in the present one to create "an invisible conveyor." These factors also help shorten production time and reduce work-in-process.

Load leveling and pull-by-subsequent-process make it possible to *rationalize operations,* i.e., make them more efficient, and to *stabilize processes,* i.e., to make them self-regulating in response to small fluctuations in production. The end result is simpler controls and greater efficiency in operations. [Editor's note: "Rationalization" refers to improvements that achieve the highest possible levels of operating efficiency; at Canon it is also used to refer to improvements involving capital investment rather than waste elimination. For example, investing in automated spot welders is a "rationalization" of the welding process.]

Many improvements in these four areas were implemented during the development of Canon's continuous flow system. Case examples from individual factories illustrate how some of the most significant improvements were developed.

Case Study 8.1 — Systematization at Factory F

This case illustrates how reductions in work-in-process and personnel were achieved when the machining process for main-body mounts was systematized.

The work center in this example uses tubing to process main-body mounts (for use in cameras such as the F-1, A-1, and AE-1).

Situation before improvements. The arrangement of machines and flow of materials before improvements is shown in Figure 8-5. Here are the most significant features:

- Only two out of six index machines were equipped with autobars (a device for feeding material automatically).
- An all-purpose machine handled any processing beyond the capacity of the one C-10RM machine.
- Handcarts were used to transport material from one installation to another.
- Workers washed the mounts by putting them in the fixtures one at a time.
- Lots were processed on a two-week cycle.
- Twelve workers were required for the operation.

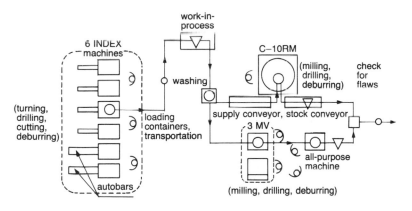

Figure 8-5 Layout and Flow of Materials Before Improvements

Situation after improvements. The state of the operation after improvements is shown in Figure 8-6. Significant points for comparison include the following:

- All six index machines were equipped with autobars.
- All processing was done on two C-10RM machines.
- Installations were linked by automatic conveyors and synchronized.
- Automatic washing equipment was introduced so washing could be done during travel.

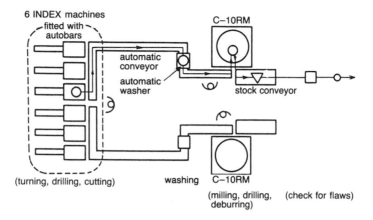

Figure 8-6 Layout and Flow of Materials After Improvements

- Processing was done in daily lots.
- Number of workers was reduced to four.

Improvement items implemented:

- C-10RM machine was converted to mixed-load processing (mounts for AE-1, A-1).
- An additional C-10RM machine installed.
- All index machines were equipped with autobars.
- Automatic washing equipment and automatic conveyors were installed.
- Setup operations were eliminated. A design change standardized processing holes for EE pin in different types of main-body mounts.
- To aid in tracing defects, a marking system indicated the number of the index machine used in processing.

Results of improvements. As a result of these improvements, work-in-process was reduced by 5,000 pieces, and personnel were reduced by 75 percent — from twelve to four.

Case Study 8.2 — Small Lot Size at Factory G

In this case small-lot processing in the caulking process was made possible by reducing the setup time from four minutes to fifteen seconds. The work center in this example performs a caulking process on parts for small items.

Situation before improvements. Setup required skill and a single setup took four minutes to complete. The actual procedure is shown in Figure 8-7.

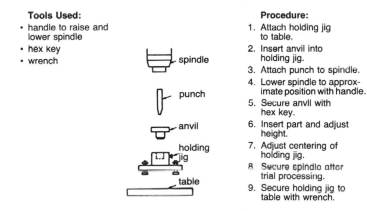

Tools Used:
- handle to raise and lower spindle
- hex key
- wrench

spindle

punch

anvil

holding jig

table

Procedure:
1. Attach holding jig to table.
2. Insert anvil into holding jig.
3. Attach punch to spindle.
4. Lower spindle to approximate position with handle.
5. Secure anvil with hex key.
6. Insert part and adjust height.
7. Adjust centering of holding jig.
8. Secure spindle after trial processing.
9. Secure holding jig to table with wrench.

Figure 8-7 Setup Procedure Before Improvements

Situation after improvements. By eliminating steps in the procedure, setup time was reduced to fifteen seconds and simplified so that anyone could perform the setup correctly. Trial runs were no longer necessary. See Figure 8-8.

Improvement items implemented:

- Height of jigs was made uniform.
- Base was redesigned to permit fixture to slide in for "one-touch setup" with self-adjusting clamps.
- Additional tools were no longer required for setup.

Results of improvements:

- Setup time was reduced from four minutes to fifteen seconds.
- Processing lot was reduced from 3,000 to 100.
- Need for trial runs was eliminated.
- Setup could be performed by anyone — quickly, correctly, and safely.

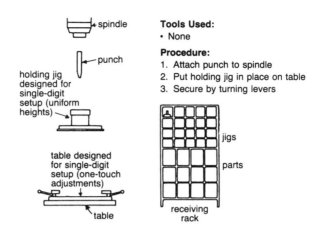

Figure 8-8 Setup Procedure After Improvements

Case Study 8.3 — Load Leveling at Factory T

At Factory T, an assembly sequencing plan resulted in leveled mixed-model assembly in two-hour cycles.

The work center in the example does final assembly for the L and J copier lines.

Situation before improvements. As shown in Figure 8-9, three groups of products in the L line (groups A, B, and C) were being produced in one-day lots (with changeover), using a single conveyor belt.

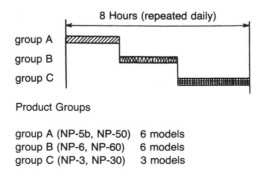

Product Groups

group A (NP-5b, NP-50)   6 models
group B (NP-6, NP-60)   6 models
group C (NP-3, NP-30)   3 models

Figure 8-9 Flow Before Improvements

Situation after improvements. The line is now used for mixed production of both the L and J product lines. One group has been added, composed of eight types of machines on the J line, making a total of four goups. The flow of products is shown in Figure 8-10. Groups A, B, and C of the L line are produced in two-hour lots at the same time as the newly added J line, thus realizing complete mixed-load processing in a single flow.

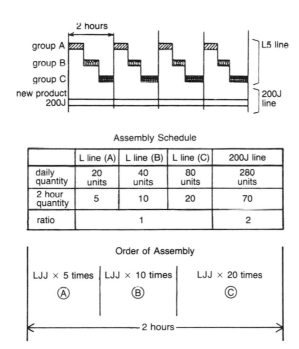

Assembly Schedule

	L line (A)	L line (B)	L line (C)	200J line
daily quantity	20 units	40 units	80 units	280 units
2 hour quantity	5	10	20	70
ratio	1			2

Order of Assembly

LJJ × 5 times	LJJ × 10 times	LJJ × 20 times
Ⓐ	Ⓑ	Ⓒ

←——————— 2 hours ———————→

Figure 8-10 Flow After Improvements

Improvement items implemented:

- Size of parts container was reduced to hold no more than the number used in one two-hour cycle.
- Parts for each unit were prepacked in sets to eliminate preparatory operations on the assembly line.
- Assembly sequencing plan was drawn up and introduced.

Results of improvements:

- Work-in-process was reduced by one day's worth.
- One-third of assembly line and 20 square meters of work center space were saved.
- Manhours in materials handling were reduced by two hours.
- One day's fluctuations in load is now held within two hours.

BASIC PRINCIPLES OF THE HIT PRODUCTION SYSTEM

1. All Production Is Handled on a Per-Day Basis

Assuring on-time delivery is an important goal of the HIT system. Delivery time is one of the pledges made between processes. When this promise cannot be kept, losses in waiting and excess production occur.

Production is handled on a per-day basis to assure delivery and prevent these kinds of losses. Per-day production means producing the output required on a particular day within that day. It means making one day the unit for completing a job. In practice, however, the unit is not always a day. Ideally, per-piece or "one-piece" production is the goal, but per-hour or per-week are also conceivable, depending upon the current management level, production configuration, and product characteristics.

Measures taken to handle quality defects are also managed on a daily basis. Under the HIT system, defective products turning up on a particular day must be taken care of within that day, while production volume is maintained.

2. Work-in-Process Is Held by the Process That Produced It

At Canon, every process is required to store the work-in-process it produces. This rule ensures that processes will feel the pinch when they generate too much work-in-process and encourages them to improve.

On the other hand, if the work center's actual capacity is

not considered and it is allowed too little work-in-process, the work center may fall behind frequently and have difficulty keeping up with the production schedule.

To prevent this, Canon sets the level of work-in-process for each work center on the basis of its actual capacity. In each work center the ratio of work-in-process value to current productivity rate is calculated, so that subsequent improvement can also be measured. The smaller the ratio, the better. The work center is also evaluated on the ratio of available space to level of work-in-process.

Figure 8-11 illustrates three typical reasons for holding work-in-process. Here are some methods for eliminating them:

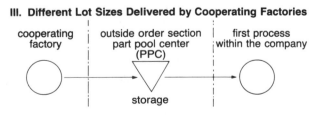

I. Differences in Processing Lot Sizes

II. Buffer Inventories in Case of Trouble

III. Different Lot Sizes Delivered by Cooperating Factories

Figure 8-11 Reasons for Holding Work-in-Process Inventories

Work-in-process caused by different processing lots.
When lot sizes differ from process to process, work-in-process can accumulate before or after any process. However, if lots for successive processes are matched, no accumulation can occur. To accomplish this, improvements must be adopted that will permit lot size reduction, such as reduction in setup time, mixed loading, and so on.

Buffer inventories ("just-in-case"). Buffer inventories accumulate when chronic problems are accepted as normal. They can be reduced by preventing trouble in the first place — by maintaining equipment and following standard operating procedures.

Work-in-process caused by gaps in lot size between the company and suppliers. At Canon, work-in-process accumulates in the Parts Pool Center (PPC) within the outside order section. The PPC inventories are caused by suppliers who are unable to fulfill Canon's internal lot-size requirements. To reduce these inventories, the outside order department staff advise suppliers on improvements that can lead to delivery in small lots.

3. Pull by Subsequent Process

"Pull by subsequent process" is a fundamental rule of the HIT production system. With it the factory moves away from the tradition of centralized control toward a simpler, self-regulating system that is responsive and flexible. In a traditional "push" system, processing orders may be issued to every work center individually. Work-in-process builds and accumulates for a variety of reasons — to create buffer inventories to cover problems, to fill a warehouse, to take advantage of machine capacity, etc. Each work center "pushes" on what it produces to be accumulated and added to at the next process.

In a pull system, work-in-process levels are reduced to the minimum; each process sends someone to take, or "pull," the exact amount of material needed from the previous process. That process then sends someone to the process before it to pull an amount equal to the amount taken. In this way, each process bases its production on the demands of the next process, and control over excess production is built in. At the same time, it becomes easier to synchronize the various processes so that there are no delays and no accumulations of work-in-process.

The system is simple and self-regulating. Whenever a process takes material from the previous process it is actually issuing a processing order at the same time. Overall production scheduling is done by an EDP system at the plant level, but only final assembly receives an actual daily production schedule. The requirements for each preceding process are automatically determined by the quantity and timing requirements given to final assembly.

4. Use Visual Control (With Actual Products as Controls)

Under the HIT production system, each work center is arranged so that control can be maintained visually, through the products themselves. First, work-in-process must be reduced to the point where abnormalities (in amount, quality, or handling) can be identified easily, simply by looking at the work-in-process. At the same time, a storage space for work-in-process is designated and the line is organized so that the quantity of work-in-process can be determined at a glance by the height and number of containers.

When the production line is set up in this manner, quick visual judgments can be made regarding trouble, process delays, or other abnormalities. The parts themselves become a check on abnormalities in the system. The scheduled production amount for the day is visually indicated on the shop floor, and delays can be detected by the accumulation of parts at one station or by comparing the level of finished goods to the scheduled amount. If the person sent to draw parts from the previous process discovers a shortage in the storage area, he informs the workers there by calling out, "Abnormality!" In this way, the problem can be addressed immediately and permanent measures can be taken to prevent its recurrence.

SELF-REGULATING PROCESS CONTROL — THE HIT CARD SYSTEM

The basis of the HIT card system used in Canon's repetitive mass-production factories is leveled production. HIT cards control and fine tune the leveling established.

Functions of HIT cards

The HIT card (Figure 8-12) automatically controls the production volume between successive processes. Because cards are used to link production between the various processes, they also control production throughout the factory. Specifically, the HIT card serves as:

- a ticket for actual articles
- an order for conveyance or delivery
- an order for processing
- a source of information for process control

Figure 8-12 Basic HIT Card

The following case study illustrates one special form of HIT card — the container HIT.

Case Study 8.4 — Container HIT at Factory C

Factory C produces a high volume of just a few models. Therefore, it is counterproductive to attach HIT cards to each parts container. There are many containers, but relatively few parts, in fairly constant amounts. For this reason, at Factory C special containers are permanently labeled with the information usually entered on HIT cards and used in place of the cards. (Photo 8-1) Each process takes containers filled with parts from the previous process in exchange for empty ones. That process then produces the amount represented by the empty containers and places the completed articles in the "store," or exchange point. The containers are in a closed circuit between the process that makes the order and the one that fulfills it. The number of containers is calculated and regulated as if the containers themselves were HIT cards.

Since work-in-process amounts are clearly defined by the containers, it is easy to tell whether an abnormality has occurred by looking at the work-in-process.

"Container for exclusive HIT use" printed on fronts of containers

Photo 8-1 Container HIT at Factory C

Basic Rules of the HIT Card System

Certain basic rules that must always be followed have been established for the HIT production system. (Figure 8-13)

Basic Principles of the HIT Production System	Rationale	Measures that will ensure reliability	Goals - Results	Remarks
Basic Principle 1 Do not send defective items to the next process	Defects are a major cost; they create disorder and disrupt the process. Give defect prevention the highest priority	When defects occur: • Machinery stops automatically (Automation) • Work stops immediately (TSS - Stop and Take Action Now) • Defects are corrected as soon as they occur (Per day handling)	• To identify and address root causes • To promote cooperation and effective defect prevention	
Basic Principle 2: Each process takes the parts it needs from the previous process	Each process knows how much it needs and when	Follow the HIT rules consistently: 1. Do not take parts without submitting HIT cards 2. Do not take more than the number of HIT cards submitted 3. Parts must always be accompanied by HIT cards	The number of parts needed by each process and when they are required is determined automatically	
Basic Principle 3 Produce no more than the quantity taken by the next process	Each process makes the quantity taken by the next process	Follow HIT rules consistently: 4. Do not produce more than the number of HIT cards received 5. Produce in the order the HIT cards are received	Overproduction and missing items are prevented and work orders are generated automatically	Synchronized production becomes possible ("constructing the invisible conveyor")
Basic Principle 4 The volume and variety of production must be leveled	To eliminate the waste that occurs when a process draws parts unevenly and forces the previous process to operate at the high end of its capacity	Use small lot sizes Use cycle times to plan production ("cycle-ization") Level the sequence of production	Since load cannot exceed capacity of the previous process, excess (reserve) capacity can be kept at the lowest limit	
Basic Principle 5 HIT cards are the means of fine-tuning production.	To respond to changes in the production schedule without changing the production setup (personnel, equipment, work-in-process) Adjustments to plan made on a daily basis so the range of fluctuation is limited and changes stay small and manageable	Leveled schedule changes Consistency as a means of fine tuning:	Fine tuning day-to-day means less need to alter production set up (personnel, equipment, work-in-process)	Application within the limits of leveling
Basic Principle 6 Stabilize and rationalize the process	To maintain a stable supply of parts and produce at the lowest cost possible	Thorough understanding and rationalization of operations Radical improvements	Standardization reduces waste, excess and unevenness in work methods and time, eliminating defective work as well as defects	

Figure 8-13 Basic Principles of the HIT Card System

Rule #1: Do not send defective items to the next process.
The identification and correction of defects constitutes a major cost and a major obstacle to production efficiency in any company.

At Canon the highest priority is given to correcting defects immediately and taking consistent measures to prevent recurrence. Wherever possible machines are designed to stop automatically when defects occur. Work also stops when anything goes wrong, so people can take action immediately. When defects are identified immediately and investigated where they occur, the root causes are more easily identified. Managers and supervisors can then work cooperatively to prevent further occurrences. All defective items are corrected or replaced on the day they occur.

Rule #2: Withdraw the parts needed from the previous process. Since every process automatically receives a "processing order" whenever parts are withdrawn from its store, it will always know how much it needs and when. For this reason, each process is responsible for withdrawing that quantity from the preceding process. That process in turn is responsible for withdrawing an equal amount from the process before it, and so on. The amounts that can be withdrawn throughout the system are determined by the number of HIT cards allowed to circulate.

Therefore:

- Do not take parts without submitting HIT cards.
- Do not take more than the number of HIT cards submitted.
- Parts must always be accompanied by HIT cards.

Rule #3: Produce no more than the quantity withdrawn by the next process. To maintain control over production — to prevent the accumulation of work-in-process or shortages — each process must produce *to order,* based on the number of HIT cards, and *in the order* in which the cards are received.

Therefore:

- Do not produce more than the number of HIT cards received.
- Always produce parts in the order HIT cards are received.

Rule #4: Level the volume and variety of production (smooth production). If a process withdraws parts unevenly (in terms of amount or timing), the previous process will be forced to operate at the high end of its capacity in people, equipment, and work-in-process. To reduce this waste, it is necessary to reduce the fluctuation, by leveling production from process to process. To achieve this:

- *Use smaller lot sizes.*
- *Use cycles to plan production.* When production is repeated in a pattern and cycles are shortened, management becomes easier all around.
- *Level the sequence of production.* This helps realize synchronized production — "the invisible conveyor."

Since process load cannot exceed the capacity of the previous process, eliminating unnecessary fluctuation in demand will allow each process to function with excess or reserve capacity at its lowest limit.

Rule #5: Use HIT cards to fine-tune production. The HIT card system allows the factory to respond quickly and effortlessly to sudden small fluctuations in demand. Since only final assembly receives a schedule for the day's production requirements, all earlier processes are expected to respond to a daily demand that fluctuates within a certain range. They simply fill the orders represented by the HIT cards, drawing on reserve capacity as the frequency of HIT card transfers increases. This capacity to adjust easily within a range of fluctuation is what is meant by fine-tuning.

When demand changes, the production plan is adjusted on a per-day basis to limit the range of fluctuation and keep changes small and manageable. This allows the fine-tuning capacity of the cards to function. As a rule, large or seasonal changes that can be anticipated are leveled to permit day-to-day reliance on fine-tuning whenever possible. For example, leveling sales volume throughout the year reduces the number of major changes in the organization of the line (i.e., re-computing the number of workers or cycle time).

Rule #6: **Stabilize and rationalize the process**. The ultimate goal is to stabilize the supply of parts and produce at the lowest cost. Through standardization and rationalization of all operations, the three Ms — waste *(muda)*, excess *(muri)*, and unevenness *(mura)* — in work methods and work time disappear. Both defective work and defective products are eliminated.

Circulation of HIT cards

The standard method of circulating HIT cards at Canon is illustrated in Figure 8-14.

At the assembly work center, the following steps are taken:

1. In assembly, when the first article is used, the assembly HIT card is removed from the container and put in the "mailbox."

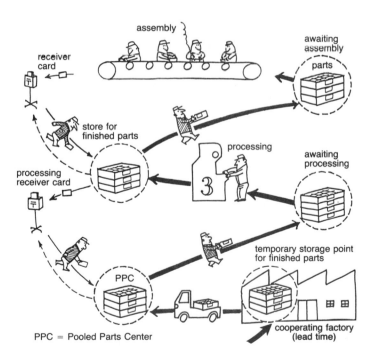

Figure 8-14 Circulation of HIT Cards

2. On a fixed cycle and at scheduled times, a worker assigned to conveyance takes the HIT cards from the assembly mailbox and goes to pick up parts from the processing
   section's finished articles store.
3. On each container he takes, he puts an assembly HIT
   card in place of the processing section's HIT card. He
   then puts the processing HIT cards he removed in the
   processing section mailbox and takes the parts back to
   assembly.

At the parts processing center, this procedure is followed:

1. At scheduled times, a conveyance worker goes to the
   store, takes the HIT cards that are in the mailbox there,
   and proceeds to the finished articles store of the previous process to get parts.
2. He puts his HIT cards in place of the ones attached to
   the containers in the finished articles store, places the
   HIT cards he removed in the mailbox, then takes the
   parts back to the deposit point for incoming stock in his
   process.
3. The parts are processed within the processing limit
   time and then put in the store to replace those that were
   taken away earlier.

Calculating the Number of HIT Cards to Issue

The number of HIT cards issued for circulation is calculated by the following method:

$$\text{Number of HIT cards} = \frac{(PT + CT) \times (\frac{\text{Daily quantity}}{8})}{QH}$$

QH is the quantity corresponding to one HIT card
(rounded off). PT or *processing limit time* is the time needed to
take the parts from the previous process plus the time needed
to process them and place them in the store. CT or *conveyance
cycle time* is the time needed to process one lot. For example,

the conveyance cycle time of a part withdrawn for processing once a day is eight hours. If it is withdrawn four times a day, the time is two hours, and so on.

Special Handling

Express cards. These function as top-priority processing orders. They are issued when trouble on the line causes missing items that create a serious obstacle to production.

Special cards. These are issued when a temporary increase in production is necessary because a subsequent process is working overtime or on holidays, or when a buffer must be built up temporarily to allow for machine maintenance time. Whenever special cards are put into the system, material and production time requirements for prior processes are also taken into account.

Odd-number cards. Occasionally defects prevent the completion of a lot ordered under a HIT card. These partial or odd-numbered lots of items are not placed in the store, but are held separately in storage as incomplete numbers. An odd-number card is used to make up the missing articles on the next lot. The completed lot is then given priority placement in the store on a first-in, first-out basis.

HIT "Scenery and Props"

The HIT production system cannot function well unless people on the production line follow the rules. Therefore, the workplace should be organized to make adherence easy. One way is to implement visual controls — so people can tell at a glance if a problem has occurred. The following "scenery and props" of the HIT system are illustrated in Photos 8-2 to 8-8.

Store. A work center's finished articles are placed in a store within the work center to serve the customer — the subsequent process. The store is set up so that the customer can withdraw parts easily:

Photo 8-2 Store

Photo 8-3 Assembly mailboxes

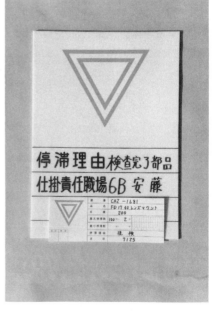

Photo 8-4 Processing mailboxes

Photo 8-5 Signboard explains
work in process accum-
ulation "reason for stor-
age: parts that have been
tested. Responsible for
work-in-process: Work
center 6B, Ando."

1. Shelves are installed for more efficient use of space.
2. Every article has an assigned space, arranged by destination and part number.
3. Addresses are displayed prominently so that anyone can read them at a glance.

Mailboxes. Mailboxes to receive the HIT cards are compartmentalized by part number or conveyance cycle to make them easy for the various work centers to use.

Signboards. Signboards are used in the work center to identify special areas, materials, or equipment, or to explain their presence in the work center (e.g., reasons for extra work-in-process).

Pushcarts. In some cases, pushcarts are used in the stores as HIT units. A card is attached and then the cart and its contents can be taken directly to the subsequent process without reloading.

Containers. In the HIT process control system, containers function as a kind of jig. Because all containers are standardized to hold a fixed number of articles, they are an indispensable tool for maintaining visual control over the level of work-in-process.

Photo 8-6 Signboards in storage point for empty containers

Photo 8-7 Store Pushcarts

Photo 8-8 Containers

Adapting the System to Different Production Requirements

The basic principles of the HIT system are uniformly applied throughout the company. Adaptations have been developed, however, to suit the type of production done in different factories.

Camera factories (high-volume production of a few models on speculation). In camera factories the number of models is relatively small and the volume of production is high. For this reason many camera factories have almost completely dedicated production lines. Moreover, there is little fluctuation in production volume, because total volume is based on sales forecast rather than orders. For these reasons, processing lots are in one-day quantities, and the lot is the unit to which one HIT card is attached. Since the daily flow of one-day lots approaches a fixed schedule, there is little need for fine-tuning. The system functions well under these conditions with the lot as the HIT card unit.

Business machine and lens factories (medium-volume, multimodel production on speculation). In copiers and lenses the number of models is large and there are great differences in production quantities among the different models. For example, in lens factories, orders for monthly production of 100,000 lenses can be mixed with small lot orders for monthly production of only a few. Furthermore, fluctuations in production volume from month to month are comparatively large. In this type of factory, the container is the unit to which the HIT card is attached. Adjustments for minor day-to-day fluctuations are self-regulating, and larger day-to-day changes in production plans can be handled by the addition or removal of a few HIT cards.

Optical instrument factories (low-volume, multimodel production on order). Production in optical instrument factories is completely different from that of cameras or business machines. Since the volume is low, with many models on order, the degree of repetition is also low and it is difficult to adapt the cards to it. The basic structure of the HIT production system remains unchanged, but in these factories the *signal system* has been developed — a computerized, regulating system that allows low-volume, multimodel production to be handled efficiently. (See pp. 147-153)

Cooperating factories. For the most part, deliveries from cooperating factories have been synchronized with processing in Canon's own factories. Some suppliers, however, do not have the capacity at present to deliver in lots as small as Canon's processing lots. These lot discrepancies take the form of work-in-process held in the Canon PPCs (parts pool centers).

To bring the continuous flow system into cooperating factories, Canon provides guidance and assistance in pushing through the improvements in manufacturing methods and equipment that have been successful at Canon.

The longer production times at cooperating factories make up a large percentage of Canon's total production time. To shorten that total time by helping to reduce production time in the cooperating factories is a continuing and important task for Canon.

STANDARDIZATION AND EVALUATION OF THE HIT SYSTEM

Establishing the "HIT Constitution"

As it was introduced, the HIT system took a variety of forms, depending on the type of production carried on in the individual factories. To make sure the system was implemented effectively throughout Canon, a list of basic principles (the "HIT constitution") was drawn up to identify the strategies considered essential in *any* production environment. The object was a simple set of standards that could be implemented immediately regardless of existing conditions. After that it would be only a matter of raising levels. The following principles were considered the most important. Each helped ensure that every work center and department would be responsible for its own work-in-process:

- Pull by subsequent process *daily* (at a minimum).
- Each process *may* handle its processing as a single lot

(i.e., may hold work-in-process) in keeping with its actual capacity.

- Handle work-in-process resulting from gaps in lot size between the company and its suppliers by setting up a pooled parts center (PPC) under the management of the outside order section.

Achieving a target is important, but maintaining and raising levels from that point on are unlikely without effective management during and after the change. Under CPS, developing the HIT system was an important goal. However, building a management system that would make it easy to implement and evaluate HIT's progress was equally important. Several important steps were taken at Canon to simplify and clarify responsibility and evaluation.

The first step was to establish the ideal work-in-process rotation time and daily pull quantity standards. Using these as a base, managers then established goals for overall improvement

	Overall Indicators	Individual Indicators
Division chief	Department work-in-process rotation time	• Change in level of daily "pull" quantities • Daily "pull" quantity rate (HIT rate)
Production section chief	Department work-in-process rotation time	• Change in time spent handling production schedule changes • Change in value of coefficient for resistance to flow • Change in time spent on startup in first process
Manufacturing section chief	Section work-in-process rotation time	• Daily quantity processing rate • Change in process integration index • Change in lead time for processing
Outside order section chief	Outside order PPC work-in-process rotation time	• Daily quantity delivery rate • Changes in lead time for ordering

Figure 8-15 Unification of Management Indicators

and planned manageable implementation levels for daily pull quantities. Once a standard and a plan were established, day-to-day work was devoted largely to identifying and overcoming obstacles to goal achievement.

Next, it was important to make sure problems would be handled by those people in the best position to solve them. As mentioned earlier, work-in-process accumulation was only allowed in departments that produced it. In this way, department chiefs who planned work-in-process reduction were only responsible for the accumulation their own departments created. Similarly, the PPC concentrated work-in-process accumulation from suppliers in one place, where it could be dealt with effectively by the outside order department staff.

How can we know whether problems are being solved effectively? Put the entire company on the same system (HIT production) and establish simple control indicators. Work-in-process rotation time is the most basic. Other basic management indicators are illustrated in Figure 8-15, divided by department responsibility.

Measuring Improvement in the HIT System

To keep the HIT production system running efficiently, all personnel must understand how their own workplace measures up against the system as a whole. Figure 8-16 illustrates a five-stage evaluation that uses as measures the ultimate goals of improvement under the HIT system. The work center's progress can be evaluated subjectively in each of the four major improvement categories — small lot size, systematization, leveling, and pull-by-subsequent-process.

To obtain quantitative data on progress, however, the following additional measures are used:

$$\text{Daily Quantity Processing Rate (daily or smaller amounts)} = \frac{\text{Parts processed daily by work center in fixed quantities}}{\text{Total parts processed by work center}} \times 100$$

$$\text{Daily quantity Delivery Rate} = \frac{\text{Daily quantity delivery accounts}}{\text{Total delivery accounts}} \times 100$$

$$\text{Integrated Processing Index} = \frac{\text{Work centers visited by parts processed by this section}}{\text{Total parts processed by this section}}$$

The *daily quantity processing rate* calculates the percentage of parts processed in fixed daily quantities. The *daily quantity delivery rate* is used by the outside order department to monitor the percentage of supplier parts that can be delivered in daily quantity or smaller lot sizes. The *integrated processing index* measures a section's progress toward sequenced processing or systematization. The goal of sequenced processing is to perform all processing of a particular part, i.e. machining, drilling, grinding, etc., in a single work center. So this index quantifies the extent to which work centers in the section are laid out according to the part processing flow rather than by type of machine. It is determined by counting how many work centers a part must pass through before it is completed.

THE SIGNAL SYSTEM

The signal system offers some of the benefits of a continuous flow system to plants with multimodel, low-volume production. It uses a computerized information system to plan and maintain daily schedules and to prevent shortages. It is called a signal system because various kinds of signals or informational cues are used to maintain control. For example, the moment a foreman realizes that external factors may prevent the day's work from being completed, he sends a "signal" to his superiors and staff departments requesting quick action to prevent any delay from carrying over into the next day.

The following case study illustrates how the signal system works in one of Canon's optical products factories.

Stage	1	2	3	4	5
Small Lot Size	Monthly quantities	Weekly quantities	Daily quantities	Hourly quantities	One piece
Process Sequencing (within Canon)	All processes separate	All processes sequenced within work center	All processes sequenced within department	All processes except assembly are sequenced	All processes from material to assembly are sequenced
Systematization Process Sequencing (Subcontracted process)	Processes subcontracted individually	No parts "ping pong"** back to the same work center after being subcontracted	Primary and secondary** processing are integrated and go to a single subcontractor	The whole parts processing job goes to a single subcontractor	The whole unit production, including assembly, goes to a single subcontractor
Leveling	Production "as it comes"	Fixed lots	Application of group technology*** to processing cycle	Integrated processing cycle	Mixed model processing
"Pull"	"Push"	HIT cards (no excess production)	Pooled Parts Center (break down outside orders into smaller lots)	Pooled Parts Center (suppliers use same lot size)	Processing in order of arrival of HIT cards

*Ping pong ball effect — When a process is completed parts go forward, but if a defect is discovered it comes back to the previous station to be corrected and is sent on again. This coming and going is like a ball in table tennis.

**Primary and secondary synthesis — In manufacturing processes there are many operations that are similar or the same. These operations should be integrated in one process naturally. Primary and secondary refer to the first and second process to be integrated.

***Group technology — When similar parts are processed together, productivity increases.

Figure 8-16 Five-Stage System Evaluation System

Case Study 8.5 — The Signal System at Factory K

Production overview. Factory K is dedicated to the production of specialized optical products in three fields: optical equipment used in TV cameras, equipment used in medical examinations, and semiconductor aligner equipment. Lots for these various products range from one to twenty units and the number of parts needed per month is 13,000 for repetitive items and 7,000 pieces for one-shot items.

Operation of the signal system. The objective of the signal system at Factory K is to assure that no shortages occur in parts shipped to assembly, by maintaining planned daily schedules in every process. It uses a computerized information (EDP) system with two primary functions: to provide daily production plans, and to maintain conditions that will permit all processing to be completed within that day. It is made up of three subsystems: a scheduling system, a material input control system, and a signal control system.

Scheduling system. Daily scheduling plans for assembly and processing are drawn up in conference by the production and marketing departments. These plans enable the plant to meet the delivery dates for each production order. (When the plans are drawn up, considerable emphasis is placed on day-to-day load leveling to help assure their achievement.)

Material input control system. This system ensures a reliable supply of materials so the line can begin processing according to the daily schedule. It combines stocking and conveyance control methods with production control boards to manage daily progress.

Signal control system. This system promotes prevention and quick recovery from delays. When scheduled processing is delayed by late materials or preparation of jigs, equipment breakdowns, or other external factors, the foreman alerts a task force of managers and personnel from staff departments to take

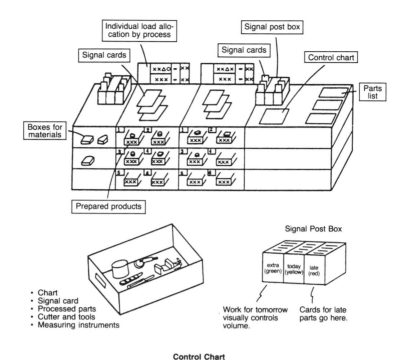

Figure 8-17 Signal System

appropriate action. A signal card is placed in a "mailbox" (Fig. 8-17) to visually indicate when parts are delayed in the previous process. The card and the current level of inventory will indicate the extent to which work must be expedited in the previous process.

Leveled Scheduling. Three techniques are used in daily schedule planning to level processing loads.

On some assembly lines at Factory K, all the products have parts in common or similar parts that are produced on the same processing lines. To plan the daily schedule for these lines, base points are first established for each five-day cycle. (See Figure 8-18) The daily schedule for assembly is drawn up by averaging the number of units to be completed for each line between these base points.

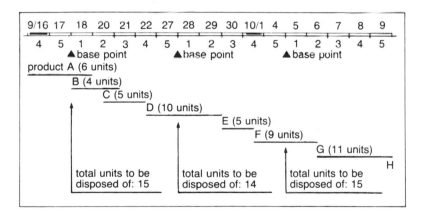

Figure 8-18 Leveling Number of Units by Establishing Five-Day Base Points

The scheduling logic of the EDP system develops the number of product units to be completed for the product orders between each base point on a parts-unit basis. Since the production volume in parts for each process is simply apportioned by working backwards from the base point in a standard order, there may be cases where the parts-processing load becomes concentrated on one machine.

In those cases, a second method is used. (See Figure 8-19) The parts-unit base point is shifted forward for processes (machines) where the load is concentrated, in order to produce a balanced processing pattern.

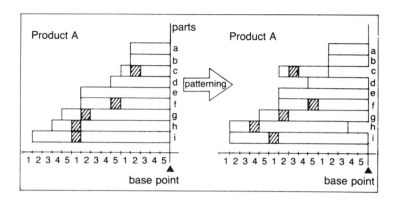

Figure 8-19 Leveling Through Pattern of Processing

Each product has an independent processing pattern and product orders change monthly, so it is impossible to guarantee leveling in every process for all products being produced. Therefore, as a third expedient, when the processing pattern for each product is established, similar parts shared by different products are scheduled for processing on the same day in the five-day cycle. This helps balance the processing loads in advance of monthly and daily scheduling.

Under the present system, the leveling objective is for 80 percent of daily loads for each process to fall within the range of the average load, plus or minus 20 percent. In the processing work centers, the overall objective is to make sure work scheduled for the day is completed that day. This is achieved primarily through scheduling and materials control and the task force approach to managing delays described above. Other measures include pushing for greater worker versatility (through training and promoting self-development), alternating processing among machines, and occasional overtime or work on holidays.

Backup for the signal system. The EDP system is used to reserve parts in stock for each order 14 days before the base point. A list is printed showing any shortage of parts, the number that are short, the lot surveyed, the processes where work is still going on, and the number of pieces in process.

Production control personnel check the list, request priority processing from the foremen of the processes where parts in question are being worked on, and schedule a completion date after making adjustments. When the scheduled completion date arrives, a confirmation of the completion of all parts — referred to as a "come up" — is produced.

Furthermore, three days before the parts are scheduled to be sent from stock, a stock file is displayed to confirm whether all parts scheduled are already in stock. Also, the parts are sent from stock in complete sets, so that any shortages can be confirmed by actually checking the parts.

These improvements in plan fulfillment and daily scheduling management ability made possible by the signal system have resulted in marked increases in the rate of fulfillment of daily schedules in both material input and processing.

WORK MANAGEMENT: WORK GUIDELINES AND PAC

Considered from the standpoint of delivery time assurance, individual operations are no different from processes: they must be organized so that work can be completed according to plan. In other words, "Plan = Results" must come true.

At Canon, work guidelines and PAC (performance analysis and control) are the means of achieving this result. Work guidelines are like contracts between workers and foremen — mutual agreements about how the work will be done. PAC is a control system that relies on performance measurements based on standard time — a scientifically established value. Both work guidelines and PAC are used to increase work center productivity through planning.

Work Guidelines

A work guideline is the optimal way to perform work operations written up in a prescribed format. An optimal operation is a safe and efficient combination of the five Ms (manpower, machines, material, methods, and measurement) under conditions defined in the work standards. Work guidelines are tools for achieving quality, cost, and delivery time objectives. The supervisor uses them to *manage* the work; the worker uses them as a guide to *do* the work.

To assure delivery time commitments, planning must include the worker level. And plans at this level must be managed so that anyone can identify a discrepancy between plan and actual results — at any time, not just at the end of the day. To achieve this, the foreman draws up a plan for each worker's individual performance. These plans (in the form of work guidelines) are posted where the workers can see them. The operation is set up so that the workers can compare the work they are actually doing with the guidelines. In this way, "Plan = Results" can become a reality and delivery time is assured. (See "The Morning Pickup System," pp.156-160, for an example of this visual control technique.)

Actual performance time is an indispensable element in drawing up work guidelines. If plans are drawn up using standard time alone, workers with lower performance can fall behind in their work while others end up empty-handed. Therefore, individual workers are timed to determine their actual performance rate. The actual times are then considered in the plans.

It is important, however, to increase work efficiency from day to day and month to month. Therefore, the foreman carries out a campaign of planned improvement and guidance, using PAC data to identify discrepancies between standard time and actual performance (measured) time. Typically, performance goals are included in workers' own improvement plans (see the three-a-month plan, Chapter 5, pp. 66-70). Once work efficiency problems are solved, the measures taken are incorporated in

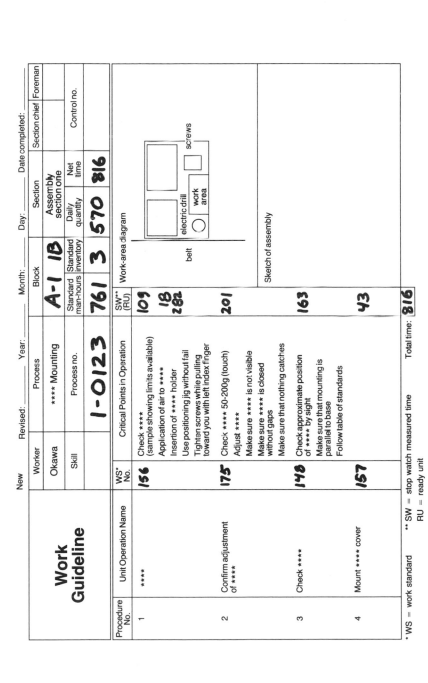

	New	Revised:	Year:	Month:	Day:	Date completed:	

Work Guideline	Worker	Process	Block	Section	Section chief	Foreman
	Okawa	**** Mounting	A-1 1B	Assembly section one		Control no.
	Skill	Process no.	Standard man-hours	Standard inventory	Daily quantity	Net time
		1-0123	761	3	570	816

Procedure No.	Unit Operation Name	WS* No.	Critical Points in Operation	SW** (RU)	Work-area diagram
1	****	156	Check **** (sample showing limits available) Application of air to **** Insertion of **** holder Use positioning jig without fail Tighten screws while pulling toward you with left index finger	109 18 282	
2	Confirm adjustment of ****	175	Check **** 50-200g (touch) Adjust **** Make sure **** is not visible Make sure **** is closed without gaps Make sure that nothing catches	201	Sketch of assembly
3	Check ****	148	Check approximate position of **** by sight Make sure that mounting is parallel to base Follow table of standards	163	
4	Mount **** cover	157		43	

Total time: 816

* WS = work standard   ** SW = stop watch measured time   Total time: 816
RU = ready unit

Figure 8-20 Work Guideline for Assembly Work Center–Sample Format

the work guidelines. In this way, operations are continually re-fined, and improvements can be shared with workers on other lines doing similar work.

A work guideline is a work plan for one cycle; it cannot be used by a worker who must perform many different operations efficiently. In such cases, *work planning charts* are used to map out all of the worker's actions for the day, from the beginning of work to quitting time.

Various types of work guidelines are used, depending on the configuration of the work:

- In continuous flow operations where one operation is performed repeatedly, work guidelines alone are sufficient. (Figure 8-20)
- In continuous flow operations where different operations are performed repeatedly in a set pattern, work guidelines are combined with work planning charts. (Figure 8-21)
- In independent operations involving different operations performed repeatedly in a set pattern, work planning charts are also drawn up. (Figure 8-22)

Case Study 8.6 — Morning Pickup System at Factory F

Visual controls can serve as a sophisticated form of work guidelines — the products themselves tell workers how they are to be handled. The following case study illustrates this type of visual control in the "morning pickup system."

The morning pickup system is a method one Canon factory developed for work-in-process reduction and visual control in parts processing. The name refers to the object of the improve-ment project: to ensure that parts picked up in the morning are finished by quitting time.

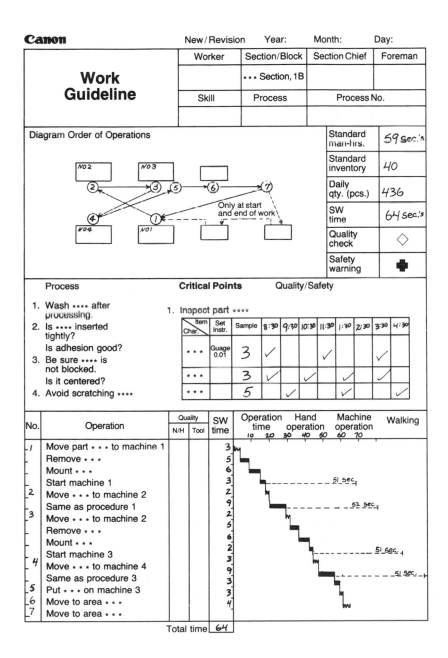

Figure 8-21 Work Guideline for Processing Work Center-Sample Format

Canon

Work Planning Chart

		New/Revised	Year:	Month:	Day:	Date completed:

	Worker	PT (min.)**	Process	Section, Block	Control No.	Section chief	Foreman
SW*(RU) / Freq.	******		****** Processing				

No.	Operations	SW*(RU)	Freq.	PT (min.)**	Operation Time (unit: hrs.) 9 A.M. 10 A.M. 11 A.M. 12 P.M. 1 P.M. 2 P.M. 3 P.M. 4 P.M. 5 P.M.	Critical points:
	Start machinery			5		Turn on electricity, air, water
	Install jig			8		
	Replenish part ******			5		
	Inspect and replenish coolant			5		
	****** operation	883	436	385		
	Conveyance of *****	1600	7X	11		
	****** inspection	2000	4X	8		
	****** inspection	1000	5X	5		
	Bring in materials	5000	2X	10		
	Remove jig			5		
	Cleanup			15		Turn off electricity, air, water
	Stop machinery			2		
	Total time:			**464**		

(Break time) 10 minutes

45 minutes

10 minutes

WG-1

* Stop Watch measured time with Ready Unit

** PT = Processing Time (total for each operation in minutes)

Figure 8-22 Work Planning Chart-Sample Format

Three basic concepts are used in operating the morning pickup system:

- Planned production and actual output for each worker are known minute by minute.
- Parts are handled in a single flow.
- Only one day's supply of parts is held.

A visual control continuously indicates the difference between planned production and actual performance. The daily scheduled amount for each worker is indicated on a panel next to an automatic counter displaying actual output. In this way the worker can tell at a glance how many parts must be processed by what time, and how many have actually been completed. (Figure 8-23) There is also a graph line showing output at 100 percent performance, so that the worker's actual performance can be measured. Looking at the counter and panel, the worker paces the work to finish the parts picked up that morning by the end of the day. The essential functions of work management are incorporated in the counter and schedule number panel. The supervisor making rounds can identify problems in the work center quickly because the progress of processing is immediately visible.

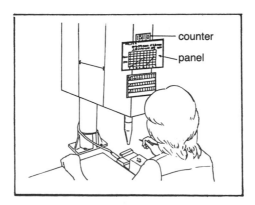

Figure 8-23 Counter and Panel Setup

The rule of holding only one day's supply of parts was introduced to reduce work-in-process and to promote an orderly, efficient work center. Work-in-process that once accumulated over three to four days was drastically reduced, and there was also a considerable saving in space.

Handling parts in a single, continuous flow requires synchronization, as in conveyor belt operations. This work center was engaged in multimodel, small-lot production using all-purpose machines. To synchronize operations was impossible unless smooth links could be made between one worker and the next. Single-flow processing was possible only after a number of improvements had been introduced, including low-cost miniconveyor belts, chutes, and specially designed fixed-number containers. (Figure 8-24)

Within three months after all these improvements were implemented, the work center saw a 69 percent increase in productivity, a 66 percent reduction of work-in-process, and a 45 percent decrease in quality defects.

Figure 8-24 Miniconveyor Setup

Performance Analysis and Control (PAC)

Broadly speaking, there are two ways to increase productivity. One is to improve the manufacturing system (renovation of equipment, improved methods, etc.). The other is to improve performance within the established manufacturing system. PAC is a system for increasing productivity through the latter approach.

Canon introduced PAC in 1971, but did not begin company-wide implementation until 1976, when CPS was introduced. Complete implementation of PAC meant putting into practice the following five principles:

1. Measure performance by scientific standards.
2. Use the guidance and interpersonal skills of super-visors rather than financial incentives.
3. Clarify areas of performance analysis and control by rank and responsibilities.
4. Report performance analytically.
5. Use a flexible, mobile work force to allocate personnel where needed to meet delivery production require-ments on daily basis.*

Canon focused its efforts on the first two items — measure-ment of performance and foreman-guided improvement ac-tivities. (An example of the latter is included in Chapter 5, Case 5.3 — "Self-Management and the Three-a-Month Plan," pp. 66-70. This section focuses on the first item — measurement of performance by scientific standards.

Setting up scientific standards. During the period of PAC implementation in 1976, Canon also made a company-wide switch from a standard time based on recorded results to a stan-dard time based on the work factor (WF) method.

To establish standard times using the WF method and then to maintain and update them was expected to take time, so a company-wide task force was established — the "ST (standard time) leveling drive." Fifteen people from all the factories cooperated in the task force effort for almost two-and-a-half

---

* Takeji Kadota, *PAC: Kōdo Seisansei no Himitsu [PAC: The Secret of High Produc-tivity]*, Japan Management Association, 1970. See also "PAC — Performance Analysis and Control," Takeji Kadota, *The Journal of Industrial Engineering*, Aug. 1968, vol. XIX, No. 8.

years. Their task was to generate the data needed to establish standard times for every assembly, machining, and surface treatment operation.

Once these data had been collected, actual calculation of the WF standard times took very little time. These were used to establish uniform standard-time levels throughout the various Canon factories. During the process of collecting the data the best methods among all the factories had been adopted, so the levels established were high.

Increasing productivity through performance. At present, three to ten people in each factory are assigned to work on standard time. They work together with the production staff members responsible for maintaining and managing standards. The standard times established, maintained, and managed by these staff members are used to measure performance in each work center or factory. (Figure 8-25)

Performance is broadly divided into areas of worker responsibility and supervisory responsibility. EDP data reports are generated on these measures in daily, weekly, and monthly form. The reports, along with other data on productivity and progress, are used by foremen and other managerial or supervisory staff. The information is used daily for guidance, for improvement activity on the line, and in cooperative activities with indirect departments.

When PAC was introduced, a temporary ban was placed on the implementation of easy work improvements at low performance levels. It was a considerable departure from established practice to limit productivity improvement efforts to control activities involving repeated observation and guidance by the supervisor. From the supervisors' standpoint, it was a disagreeable task.

Nonetheless, waste elimination attributable to PAC accounted for about 30 percent of the results in the first year of the waste elimination drive; in the following year it accounted for almost 40 percent.

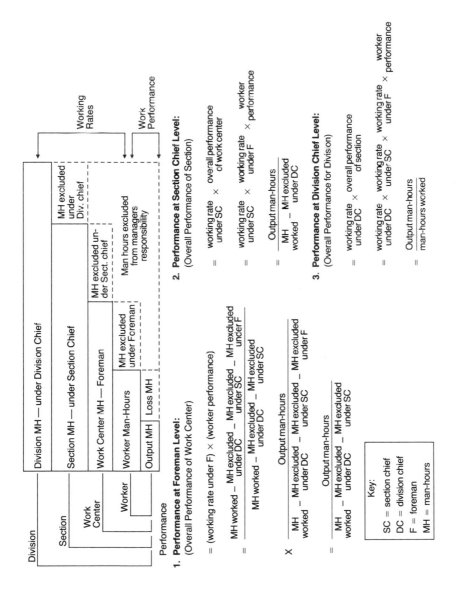

Figure 8-25 Measuring Performance for Each Work Level

PAC increased performance at Canon (its initial goal) and is now being applied to maintain performance at high levels. Changes in productivity after the introduction of PAC are shown in Figure 8-26. Within three years, productivity increased 150 percent and then stabilized as a result of increased performance alone. This does not include the effects of work improvements (overall performance). Once performance had been raised, the ban on work improvements was lifted and work improvement suggestion activity flourished.

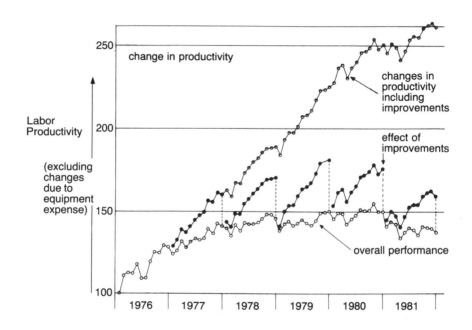

Figure 8-26   Increase in Productivity After the Introduction to PAC

Currently, loss analysis — the overall evaluation of productivity described in Chapter 7 — is practiced regularly in every work center. Analytical, planning-centered efforts to increase productivity through performance continue to accompany Canon's efforts to increase productivity through improvements in manufacturing methods.

OVERALL RESULTS OF THE HIT PRODUCTION SYSTEM

Figure 8-27 shows the reduction in work-in-process throughout Canon after the introduction of the HIT system. Figure 8-28 summarizes the overall effect of the HIT production system.

Canon experienced a broad range of positive benefits as a result of implementing HIT. Work-in-process reduction increased Canon's ability to respond to change — to meet market demands and fit in special orders and trial production without excess inventory. It also dramatically reduced costs related to managing work-in-process, because less space, time, and personnel were needed for product storage, handling, and accounting. It paved the way for synchronized production by simplifying production control processes through the self-regulating, fine-tuning functions of the HIT cards.

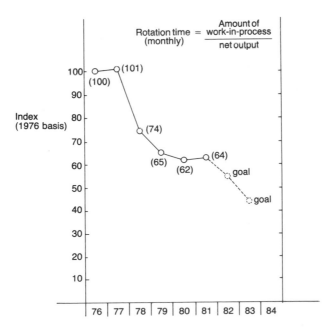

Figure 8-27 Reductions in Work-in-Process Rotation Time

Results of Implementing HIT
• Increased efficiency of capital (interest burden, availability of funds, expenditures on land and factory space)
• Less time spent managing work-in-process (managing products, storage, accounting and trouble-shooting, i.e. filling shortages, etc.)
• Greater reliability in planning (less time between setting and achieving target; fewer, less critical plan changes)
• Increased ability to respond to change (meeting market demands; fitting in special orders and trial production without excess inventory)
• Implementation of visual control (understanding which processes create bottlenecks; learning to use hidden strengths and eliminating latent weaknesses)
• Increased managerial productivity (expanded scope of responsibility; ability to process more items)
• Production control processes reduced through synchronized production (control through self-regulating and fine-tuning functions of HIT cards)
• Thorough waste elimination taught us: Overproduction is the greatest waste – waste that creates waste and hides problems. Quantify and measure all forms of waste – cost reduction is the true goal. Use economic judgment – with cost reduction as the standard, find and use hidden strengths, find and eliminate hidden expenses. Promote a flexible workforce – reduce waste → redistribute work → reduce labor costs by the man instead of the manhour. Systematize line configurations – through multiprocessing, multi-tasking, and cycling. Adopt practical principles of equipment use and design – automation; work rate vs. operating rate; avoid excess capacity (and overproduction) by using small, low speed machines; use mistake-proofing devices. Improve equipment – ONLY after work improvements have been thoroughly investigated and implemented. Reduce labor costs – by initial planning based on small numbers, through feedback after improvements. Use line stop for quality control and defect prevention – TSS – "take action now," stopping at fixed positions.

Figure 8-28 Overall Benefits of the HIT System

Keys to Fundamental Improvement

*Overproduction is the most complex and
challenging form of waste. Reducing work-in-process,
however, provides solutions to many other problems.
It promotes a work environment where human
resources can be fully developed and used.*

Canon's vigorous waste elimination efforts verified
and reinforced several important principles of constant
improvement. Cost reduction cannot be successful unless
all forms of waste are quantified and measured. And, once
everyone is accustomed to looking at waste in measurable
terms, it becomes obvious that overproduction is the greatest form of waste, creating, sustaining, and hiding many
other forms of waste.

Using waste elimination as the standard teaches
people to be practical and sensible in the use of important
resources: labor, equipment, and space. To reduce wasted
man-hours, CPS promoted the development of a flexible
work force and flexible multiprocessing, multitasking
work centers. To reduce waste in equipment, CPS pur-
chased and used equipment on the basis of production
requirements rather than capacity; equipment improvements were only made after work improvement opportunities had been thoroughly investigated.

Managing overall improvement also became more
efficient. Reliability in planning increased, as training and
continuous practice resulted in fewer, less critical plan
changes and reduced the time between setting and achieving targets. This focus on disciplined planning had another

important side effect: managers became more productive. Their responsibilities increased under CPS, but the emphasis on information sharing, planning, and mutual instruction increased managers' ability to handle more challenges more efficiently.

This effect on the software side of improvement has only reinforced Canon's conviction that using human resources effectively is one of the most reliable approaches to successful waste elimination. Careful observation and planning, concentration of effort, and quick action to keep promises and prevent problems — these are the methods Canon workers use to reduce costs by becoming more effective and responsible as individuals.

9
The Role of CPS in Quality Assurance at Canon

Canon's total quality assurance (QA) system spans all departments from development and production to marketing. In addition to quality standards, the system gives each department a role in the quality effort and establishes reporting channels to ensure effective implementation and adherence to standards.

To achieve quality — to actually reduce defects to zero — requires focused improvement at the source of defects, where work is performed. Canon's QA system and regulations provide a framework for quality assurance, but its defect reduction activities make that framework work, like the muscles and ligaments that hold a skeleton together. This chapter describes some of the many ways in which defect reduction is promoted by CPS.

ESTABLISHING QUALITY ASSURANCE OFFICES

In 1976, quality assurance offices were set up in each factory to promote the drive for zero defects. In the beginning, representatives from every factory met once a week at company headquarters to discuss how QA offices should promote the drive. In those meetings a simple analogy was developed to explain two important roles in quality assurance: "fire (defect) prevention" and "fire (defect) detection." (See Figure 9-1)

Fire Department (watchman)	QA Department (QA person)
Watchtower →	Measuring equipment
Designing watchtower and observation equipment →	Investigation, study, and development of measuring equipment
Maintaining watchtower and equipment →	Correcting measuring instruments
Looking out over the town →	Monitoring processes according to a plan
Occurrence of fire →	Occurrence of defects
Which block in which street? →	Which procedure in which process?
Construction of the burning building →	Design of process where defect occurred
Fire hydrants, dangerous materials, population density →	Information concerning context of the process
Wind direction →	Work environment
Straw roof 100 meters away →	Other processes, other production lines
Report fire to the station →	Send information about defects to relevant departments
Fire fighters →	Technological and manufacturing departments
Fire prevention activity →	Countermeasures against defects

Figure 9-1 Fire Department and QA Department

THE ROLE OF QUALITY ASSURANCE AT CANON

"Fire (Defect) Prevention"

Prevention is the most important role of the fire department. To prevent fire losses, the fire department provides fire-safety equipment and establishes regulations. It conducts on-the-spot inspections and instruction in fire safety, e.g., safe handling of flame, storage of combustible or explosive materials, escape routes and exits, etc. In the same way, the quality assurance office must prevent defects from occurring. It makes certain products are made properly from the start, by setting standards, ensuring adherence, and by testing.

"Fire (Defect) Detection"

Ideally, quality is built into every process so that everything sent to the next process is 100 percent free of defects. When defects do occur, however, the quality assurance office must make certain they are not overlooked and that measures are taken to prevent them in the future.

This is the role of the watchman in the fire watchtower. Always alert, he makes sure his viewing equipment is the best for his purpose and sees that it is carefully stored and maintained. He uses this equipment to keep watch over the town from the tower. When fire breaks out, he tells the fire station precisely where it is located. He also communicates other information needed to put out the fire effectively: the construction of the building, location of fire hydrants, presence of dangerous materials, population density, wind direction, temperature and humidity, and whether nearby buildings are endangered by flying sparks. With this information, firefighters can rush to the scene with the proper fire-fighting equipment.

If the fire gets out of hand, the firemen request assistance from neighboring fire departments. After the fire is out, an investigation into its cause is held with all concerned parties, and measures are taken to prevent similar fires from breaking out in the future.

In the same way, the quality assurance office keeps watch over the production processes, using accurate measuring instruments to gather data constantly on the occurrence of defects. This information helps production and technical departments take countermeasures against defects, and facilitates prompt and careful investigation into their causes.

QUALITY ASSURANCE FLOW CHARTS

To fulfill these two roles, the quality assurance office uses the *quality assurance flow chart,* a planning tool designed to promote quality from the earliest stages of production.

The QA flow chart breaks down the manufacturing method into detailed quality characteristics for individual components. These characteristics are expressed in the form of values. (See Figure 9-2) The flow charts also indicate the kind of sampling needed in each manufacturing process to assure high quality, which measuring tools to use and how, and where to record results.

Canon To: _____   Date drawn up: 3/16/78
Drawn up by: Kikuchi

QA Flow Chart   Exposure-counter base-plate:

Manufacturing process	Worker, Machine, Standard specifications	Quality characteristics	Characteristic value	Measure	Sampling	Record
Caulking of exposure-counter base	○ Work standard M-015-037	Caulking force	1200 g-cm (1 minute)	Torque gauge (left, 2kg-cm)	$n = \dfrac{1}{1000}$	x – Rs
	○ Highspin	Inclination of spindle tip	0.02 mm or below	Projector 20 times	$n = \dfrac{3}{1000}$	Checksheet
	○ Work center A (Sato)	Caulking defect projection	0.1 mm or below	Dial gauge	$n = \dfrac{5}{1000}$	Data sheet
		Flatness of base plate	0.1 mm or below	Dial gauge	$n = \dfrac{5}{1000}$	$\bar{x}$ – R
⟡ Standard area		All above characteristics			First products $n = 5$	Data sheet
• Measure- ment area						

Figure 9-2 QA Flow Chart

In this flow chart, the column headings can be defined as follows (reading from left to right):

- A *manufacturing process* is any operation that produces a physical or chemical change in the processed item. That change is measured to assure that quality has been built in. Operations such as washing, transportation, and storage are also treated as processes that have an influence on quality. In actual practice, this column is filled out using the assembly flow chart, the process chart, and other planning documents.

- The *worker* is the person taking the measurements. Depending on the degree of difficulty, measuring is assigned to direct or indirect workers, the foreman, or QA personnel.
- *Machine* refers to the specific piece of equipment, machinery, or jig used in the manufacturing process.
- *Standard specifications* refers to the work standard or process-chart entry number.
- *Quality characteristics* are the items or areas subject to quality evaluation.
- *Characteristic value* refers to the value of the quality characteristic — a significant difference that can be measured — expressed in the unit of measurement, e.g., centimeters (cm), grams (g), seconds (S), temperature (C), or ratio (%). Subjective expressions, such as "item not good" or "failed," are not used.
- *Measure* refers to the measuring instrument and method. Since the characteristic value is quantified, measuring instruments that express a numerical result are required. The "by eye" method is not used.
- *Sampling* frequency depends on the stability of the particular process (i.e., the rate of inconsistencies, defects, etc.). Inspection of first products is distinguished from regular periodic inspection.
- *Record* refers to the name of the register or control chart where data are to be recorded. The use of control charts (especially $\overline{X}$-R) is actively encouraged to obtain quantitative information on quality.

Development and Distribution of Flow Charts

QA flow charts are a primary means of assuring quality at Canon, so they are developed early. During the product development stage, the QA office gathers information on quality characteristics and researches appropriate measurement techniques. When the flow chart is complete, the factory QA office verifies which instruments, measurement, and recording methods will be used.

QA flow charts are distributed in a manner that ensures effective implementation. Before distribution, a chart may be tested and adjusted in trial runs; each provision is thoroughly explained so everyone involved understands what is required. Flow chart provisions are routinely incorporated in the work standards for every operation so workers understand and adhere to them. Finally, a "QA information network" ensures that data collected through the flow charts are sent to departments responsible for dealing with them.

The QA network puts quality information in a form that can be understood by people at all levels and functions and raises awareness concerning control of critical quality characteristics. Information is circulated regularly from the QA office in each factory to the engineering department and the plant management office. Through those offices it goes down to the shop floor in the form of specific feedback. Routine data on a processed part are circulated on a fixed schedule, sometimes monthly. Data on a rejected part are circulated quickly, depending on the degree of emergency represented.

QUALITY ASSURANCE INSPECTIONS

Quality standards and checking procedures are established through the flow chart. The next important step is to make sure the data gathered are used effectively — to make sure quality is built in continuously. The QA inspection answers three important questions:

1. Are established procedures followed?
2. Are measures taken to prevent the recurrence of defects?
3. Are standards revised once improvements have been made?

These questions are answered by conducting inspections in the following four areas:

Process being inspected		Overall evaluation	QA Flow chart adherence rate: ___%	Rate of drafting work standards ___%
Section	Foreman		Follow-up rate: ___% Correction rate: ___%	Rate of adherence to work standards: ___%

1. QAFC
(Quality Assurance Flow Chart)

1. Check
   - ☐ Checked according to flow chart?
   - ☐ Does the number of samples match?
   - ☐☐
   - ....
2. Data
   - ☐ Are characteristics on the FC gathered?
   - ☐ Does the number of samples match?
   - ☐
   - ....
3. Limit Samples
   - ☐ Are there photographic samples?
   - ☐ Are there limit samples?
   - ☐☐☐
   - ....
4. Measuring Instruments
   - ☐ Are specified measuring instruments present?
   - ☐ Are they used correctly?
   - ☐☐
   - ....
5. Measures and Countermeasures
   - ☐ Is there data concerning abnormalities?
   - ☐

2. Work Standards

1. Adherence (Standards)
   - ☐ Are work standards established?
   - ☐ Are standards established for processing conditions?
   - ☐☐
2. Adherence (Operations)
   - ☐ Are operations carried out according to standard procedures?
   - ☐ Is the process stable (no changes)?
   - ☐
   - ....
3. Controls Management
   - ☐ Are there stamps for study and approval?
   - ☐ Are there control numbers?

3. Management of measured values by foreman

1. Management (control activities)
   - ☐ Are the specified characteristics measured?
   - ☐ Are they displayed?
   - ☐☐☐
   - ....
2. Measures and Countermeasures indicated
   - ☐ Are abnormalities indicated with red marks?
   - ☐ Are persistent abnormalities eliminated?
   - ☐☐

4. Preventive Maintenance

1. Correction
   - ☐ Are measuring instruments corrected? (presence/absence of labels and correction marks)
   - ☐ Are time limits observed?
   - ☐☐☐

How to Conduct Inspection:

Before inspection:
1. Identify relevant standards.
2. Study last inspection report/replies.

5. Other

1. Items indicated on last inspection.
   - ☐ Are deficiencies being followed up?
2. TSS ("Take Action Now")
   - ☐ Are TSS claim measures and countermeasures adhered to?
3. Classification
   - ☐ Is the classification of parts and products handled well?
4. Safety and health equipment
   - ☐ Is a control system in place?
   - ☐ Is it followed?
5. Packing
   - ☐ Is the proper packing used?

During inspection:
3. Ascertain facts carefully.
4. Do not comment on work directly.

Key:

QAFC adherence rate = QAFC instructions followed ÷ QAFC instructions

Rate of follow-up = improvements ÷ deficiencies indicated

Correction rate = corrected instruments ÷ measuring instruments

Rate of drafting work standards = standards ÷ operations

Rate of adherence to work standards = performed to standard ÷ standards

Figure 9-3 Process Inspection Checklist-Sample Format

Work standards. Work standards for every process indicate the conditions and procedures required to produce products without defects. They also identify which points are critical. On-the-spot inspections are made while work is in progress to determine whether work standards are followed.

QA flow charts. A second purpose of inspection is to determine whether items on the QA flow chart are implemented in the prescribed manner. In some operations, however, it is difficult to get a numerical reading on quality characteristics — operations using adhesives or lubricants, for example. In spite of such limitations, rigorous inspections of the amount and place of application, storage methods, period of effectiveness, and so on, are carried out in these areas.

Measured values control. As part of measured values control, the QA office indicates to each work center which quality characteristics should be managed by $\bar{X}$-R control charts and how they should be used. Control charts are not simply posted: their function as controls must be fully realized. This means that defects are investigated as soon as they occur, and steps taken to prevent their recurrence. Inspections in this area confirm that prompt follow-up procedures are being followed.

Measuring instrument control. Correctly adjusted measuring instruments are essential to assure quality. For this reason, only instruments corrected and certified periodically by measurement-instrument control personnel are authorized for use. Inspections conducted in this area make certain that instruments overdue for correction are not used and that instruments are stored properly.

Inspections in these four areas are carried out using the Process Inspection Checklist (Figure 9-3), the Process Inspection Report (Figure 9-4), and the Process Inspection Reply Sheet (Figure 9-5).

Fukushima
Factory

QA Section   **Process Inspection Report**   Issue No. 81028

	Time of Inspection: May 13-14, 1981	approval	inspectors
Process: A-1, 3rd assembly section, 4B, Foreman: Yasuda		Imai	Kinoshita Tanabe

Overall Evaluation

This time inspection of the entire process in A-1, from 1B to 4B was completed. The impression remains that . . . .

Inspection Results
1. QA flowchart
   (a) The amount of oil was not checked for five items.
   (b) The time limit for measuring instruments was exceeded in three cases. Please correct this to follow QA measurement control G.
2. Work Standards
   (a) No work standards. . . .2 items
   (b) Work standards not followed. . . .9 items

Figure 9-4 Process Inspection Report — Sample Format

	section chief Nakamura	foreman Yasuda
Process Inspection Reply		

	1.QA Flowchart 2.Work standards	3. Management of measured values by foreman		

item	Measures and countermeasures	(daily schedule may be shown in a diagram)	daily schedule	person re-sponsible
1.	(a) Implement oil quantity CH.		5/20	Abe
	(b) Adjust for exceeding limit on measuring instruments according to QA measurement control G.		6/8	Abe
2.	(a) Draw up standards for back cover, oiling latch, strobe operation CH.			
	(b) Standard No. 105			
	• 502 pinhole inspection 45 degrees. Implement CH from right.		6/1	Yasuda
	•			

Figure 9-5 Process Inspection Reply — Sample Format

EFFECTIVE APPLICATION OF
PRODUCTION TECHNOLOGY

In 1960, Canon became the first company to mass-produce high-quality cameras using conveyor assembly operations. Since then Canon has consistently led the field in manufacturing technology. First came high-precision plastic technology and the automated production of the AE-1 camera (which started the electronic camera revolution). More recently, Canon introduced a flexible production system for mixed-model business machine production and the submicron technology used in the manufacture of semiconductors.

When Canon looked back over the process of development, however, it was clear that time and money had often been wasted, because company scientists did not always make use of each other's expertise.

For example, the production department routinely studies product development plans. It decides which processes and production methods are needed to produce a new product, in stable quantities, at an appropriate cost, and at the required level of quality. Once optimal manufacturing conditions are defined, directions are given to the actual manufacturing staff. In the past, however, workload imbalances sometimes forced departments to juggle technological staff assignments for personnel ranging from experienced veterans to new employees. When personnel changed, the directions often changed as well. Mixups like this, occurring repeatedly in some departments, led to waste in startup and defects. Diversification within product groups and the creation of additional factories had increased the complexity of the company's organization. Information exchange between departments and between factories became increasingly difficult.

CPS brought to the firm a number of activities designed to improve the transfer and application of production technology know-how. The most significant were the development of production technology standards, to prevent defects and ensure reliability of manufacturing conditions, and the production technology information system, a computerized documentary information reference system.

Production Technology Standards

Canon's production technology standards help build quality into manufacturing when processes, procedures, and manufacturing conditions are first set up. Based on the process capacity of existing processes, they reduce the tendency to rely on undefined "experience" or intuition in judging the requirements.

For example, manufacturing conditions affecting quality in soldering include the temperature and shape of the soldering iron tip, the duration of solder application, the type of solder, and the ingredients in the flux. Production technology standards related to soldering explain the cause-and-effect relationships between these factors and soldering quality characteristics — i.e., how and why certain conditions produce certain results. The standards are organized for general use and supported by technological data.

Production technology standards help prevent defects that occur during manufacturing. They also help ensure reliability in manufacturing conditions related to defects that may only show up after manufacture.

There are two approaches to standardization: the deductive and the inductive methods. Recently, the inductive method has been prompted by rapid increases in the tempo of technological innovation and the need for efficient workloads in experimentation related to standards. Laboratory experimentation is limited and greater reliance is placed on data and methods that can be gathered on the line by foremen and workers. Typically, the process of standardization follows this path:

- A standard is inferred from data observed and collected at the level of operations.
- The proposed standard is tested and adjusted according to results observed.
- The standard is adopted when results are stable and reliability is established.

Figure 9-6 Sample Manufacturing Technology Information Abstract

Production Technology Information System

In a new technology, the most important information almost always starts out as isolated data pertaining to an extremely narrow field. All data that have technological value must be collected and kept for future standardization, even though at this stage they may be fragmentary and unsuitable for broader application. To lay the groundwork for standardization, it is essential to have a system that can manage and retrieve the company's technological know-how.

Canon's technological information system is based on the technical documentary information reference method. Information related to production technology developed throughout the company is systematically collected, edited, and published in abstract form as part of a company-wide information service. (See Figure 9-6) This service provides a daily means of production technology transfer and exchange between factories and departments. At the same time it provides the data base for future standardization.

To achieve more efficient circulation and application of information, the service has been converted to an EDP on-line system. Figure 9-7 shows the operational flow of the new information system.

CAMPAIGN TO RAISE PROCESS CAPABILITY

Objectives of Increased Process Capability in Camera Production

Produce smaller, lighter, multifunctional products. Since the introduction of CPS, Canon has been involved in a significant effort to increase process capability, partly in response to the demand for high-precision parts. For example, ten years ago, smaller and lighter single-lens reflex cameras were introduced, and more recently automatic functions. One after the other automatic exposure, film advance, rewind, and

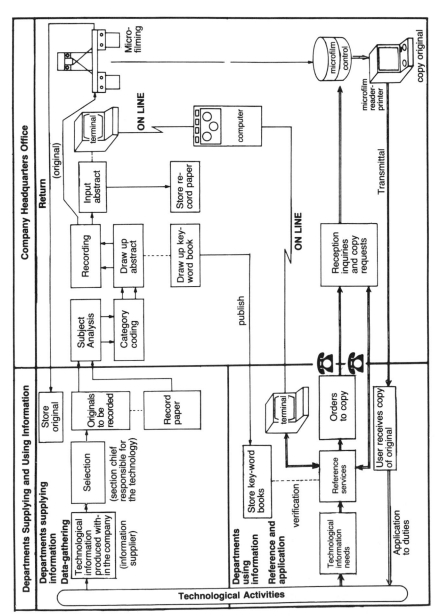

Figure 9-7 Operational Flow of Production Technology Information System

focus were added. Raising process capability became essential to meet this growing demand for high-precision parts.

Eliminate adjustment in assembly operations. Assembly is the most costly operation in camera manufacturing in terms of labor, and adjustment operations that depend heavily on human judgment are an important part of assembly. To eliminate these operations, it is first necessary to increase process capability.

Reduce defects in assembly and increase product reliability. As process capability in parts processing increases, defects decrease in assembly, adjustment operations are reduced, and work can be simplified. Since simpler work will also result in fewer mistakes than more complex operations, defects are reduced even more.

Furthermore, feedback on quality characteristics helps in estimating a product's failure rate before it is marketed. This kind of feedback can only increase product reliability.

Reduce startup time for new products. Startup time for new products can be significantly reduced, once process capability is brought under control on equipment processing current parts. By comparing new product parts to those currently being processed, parts that can be produced with existing equipment are easily distinguished from those that require new equipment or techniques. Once these distinctions are made, the points to watch for become clear and startup time can be reduced.

Process Capability Point System
Stimulates Involvement and Competition

Many people find values like "Cp" or "Cpk" (indicating process capability) hard to understand even when they are shown directly how the values are derived. That is one reason CPS introduced a simple evaluation system using points. Process capability is indicated by point scores from 60 (failing) to 80 (passing) up to 100 points, so anyone can tell at a glance how to evaluate the score. The points are based on the five to ten quality characteristics designated for each work center on the QA

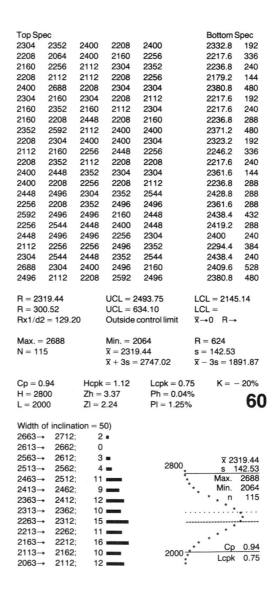

Top Spec					Bottom Spec	
2304	2352	2400	2208	2400	2332.8	192
2208	2064	2400	2160	2256	2217.6	336
2160	2256	2112	2304	2352	2236.8	240
2208	2112	2112	2208	2256	2179.2	144
2400	2688	2208	2304	2304	2380.8	480
2304	2160	2304	2208	2112	2217.6	192
2160	2352	2160	2112	2304	2217.6	240
2160	2208	2448	2208	2160	2236.8	288
2352	2592	2112	2400	2400	2371.2	480
2208	2304	2400	2400	2304	2323.2	192
2112	2160	2256	2448	2256	2246.2	336
2208	2352	2112	2208	2208	2217.6	240
2400	2448	2352	2304	2304	2361.6	144
2400	2208	2256	2208	2112	2236.8	288
2448	2496	2304	2352	2544	2428.8	288
2256	2208	2352	2496	2496	2361.6	288
2592	2496	2496	2160	2448	2438.4	432
2256	2544	2448	2400	2448	2419.2	288
2448	2496	2496	2256	2304	2400	240
2112	2256	2256	2496	2352	2294.4	384
2304	2544	2448	2352	2544	2438.4	240
2688	2304	2400	2496	2160	2409.6	528
2496	2112	2208	2592	2496	2380.8	480

R = 2319.44   UCL = 2493.75   LCL = 2145.14
R = 300.52   UCL = 634.10   LCL =
Rx1/d2 = 129.20   Outside control limit $\bar{x} \to 0$ R$\to$

Max. = 2688   Min. = 2064   R = 624
N = 115 $\bar{x}$ = 2319.44   s = 142.53
                   $\bar{x}$ + 3s = 2747.02 $\bar{x}$ − 3s = 1891.87

Cp = 0.94   Hcpk = 1.12   Lcpk = 0.75   K = − 20%
H = 2800   Zh = 3.37   Ph = 0.04%
L = 2000   Zl = 2.24   Pl = 1.25%   **60**

Width of inclination = 50)

2663→	2712;	2 ▪
2613→	2662;	0
2563→	2612;	3 ▪
2513→	2562;	4 ▪
2463→	2512;	11 ▬
2413→	2462;	9 ▬
2363→	2412;	12 ▬
2313→	2362;	10 ▬
2263→	2312;	15 ▬
2213→	2262;	11 ▬
2163→	2212;	16 ▬
2113→	2162;	10 ▬
2063→	2112;	12 ▬

2800

$\bar{x}$ 2319.44
s   142.53
Max. 2688
Min. 2064
n   115

2000

Cp 0.94
Lcpk 0.75

Figure 9-8 Evaluation Points Calculated on Microcomputer

flow charts. Evaluation points are calculated rapidly, using a microcomputer. (See Figure 9-8) The methods for evaluating process capability include frequency distribution, normal distribution, and $\overline{X}$-R control charts. Most of the data used in point evaluation are based on $\overline{X}$-R control charts. The data are analyzed and then scored according to the process capability judging standards. (Figure 9-9) Total evaluation points are used to promote competition between work centers in the campaign to increase process capability.

Using this system has emphasized the importance of process capability and promoted activity to raise it. Many factories use the point system to help fulfill the stricter precision requirements in their product parts. For example, one factory's continuing goal is to raise characteristics worth 70 points at the beginning of the year to 90 points by the end of each year. Characteristics reaching 100 points by the end of the year have clear, standardized control points and their high level is judged to be maintainable. At this point, the characteristics are eliminated as objectives in the point system, and the next year characteristics with lower scores are put in their place. This process is repeated year after year. Furthermore, when new equipment is introduced, tests are conducted with the strict requirement that a score of 80 or more points be achieved.

THE TSS DRIVE — (STOP AND TAKE ACTION NOW!)

TSS is an important management technique for assuring product quality at Canon. TSS stands for the Japanese words *tomete* (stop), *sugu* (right away), and *shochi o toru* (take measures). When a problem occurs:

> T — stop the line or machinery
> S — right away and
> S — take measures to correct it.

Stopping the line is discouraged in a traditional mass-production system using conveyors. Except in extreme cir-

Judgment Items: Defect rate P (*Cpk*)

Note: *Cpk* is an index of process capacity based on evaluation of the inclination of $\bar{x}$.

Judgment Standards:

	No. of points
P 0.03% ---→ (1.17 ≤ *Cpk*)→	100
0.03% < P ≤ 0.13 ---→ (1.00 ≤ *Cpk* < 1.17)→	80
0.13 < P ≤ 5 ---→ (0.55 ≤ *Cpk* < 1.00)→	60
5 < P ≤ 15 ---→ (0.35 ≤ *Cpk* < 0.55)→	40
15 < P ≤ 50 ---→ (0 ≤ *Cpk* < 0.35)→	20
50 < P ---→ (*Cpk* < 0)→	0

Method of Calculation: 1. **Determination of defect rate P: read by applying Z to expression of regular distribution.**

$$Z_H = \frac{H - \bar{x}}{\sigma} \quad \text{and} \quad Z_L = \frac{\bar{x} - L}{\sigma}$$

$$Cpk = (1 - K)\frac{H - L}{6\sigma} \quad \text{(K is the degree of inclination of } \bar{x}.\text{)}$$

$$= \frac{H - \bar{x}}{3\sigma} \text{ or } \frac{\bar{x} - L}{3\sigma} \quad \text{(The smaller value is taken.)}$$

Note: Since the defect rate is the sum of both sides, it does not correspond perfectly to *Cpk*.

2. **Determination of K: K expresses the degree of inclination of $\bar{x}$ from the center of the specification.**

$$K = \frac{\bar{x} - \frac{H + L}{2}}{\frac{H - L}{2}} \times 100 = \frac{2\bar{x} - H - L}{H - L} \times 100 = \frac{A}{B} \times 100$$

Note: Not expressed in the event of one-sided specifications except for exceptional cases.

Example of Scoring: Manufacturing Section 1   Work Center 1

No.	Quality characteristics	Points
1	15 ± 0.03	60
2	28 ± 0.03	80
:	:	:
10	4 ± 0.03	80
	Total	840

$$\frac{\text{Evaluation}}{\text{points}} = \frac{\text{Total points}}{\text{Number of characteristics}} = \frac{840}{10} = 84.0$$

The evaluation score for Manufacturing Section 1, Work Center 1 for January, 1979 is 84.0 points.

Figure 9-9 Judging and Scoring Process Capacity

cumstances, there is a tendency to rig up hasty emergency solutions to problems as they occur. However, this is a lot like the old folk method of beating the ground to get rid of moles. A defect may be stamped out on that particular occasion, but it invariably pops up again later. Everyone is kept working, but in the long run the same ineffective remedies are repeated over and over, with no positive result. TSS is one of the ideas Canon workers came up with to pinpoint and overcome problems of this kind.

Traditionally, the worker has no authority to stop the line — TSS gives the worker that authority. TSS is part of a quality control program to prevent problems at their source rather than trying to catch them after they have occurred. It cannot function well, however, unless factory management understands and supports the program completely.

"Stop the Line So It Won't Have to Be Stopped"

Stopping the line when problems or abnormalities occur is not the ultimate goal of TSS. On the contrary, the line is stopped the moment a problem occurs in order to *eliminate* problems, so that in the future the line will not have to be stopped.

In the context of TSS an abnormality occurs whenever:

- defects occur within the process
- parts from the previous process are defective
- guidelines cannot be followed
- there are no parts
- work in one's own process falls behind

In other words, whenever work cannot be done according to work standards or production requirements, the worker stops the line. Supervisors and managers must do their utmost, however, to anticipate or bring problems to the surface beforehand, so the line need not be stopped.

"Take Corrective Action Right Away"

If a problem *does* occur, the worker stops the line and

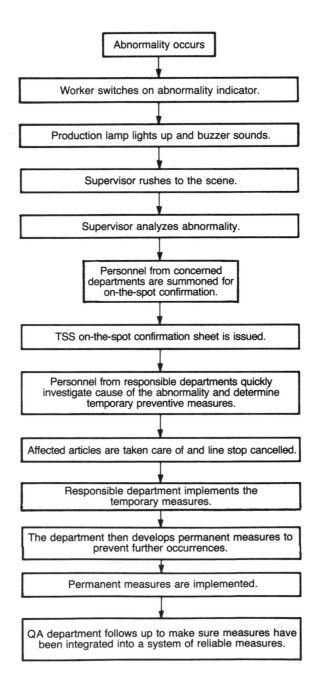

Figure 9-10 Flow of the TSS System

switches on the abnormality indicator; the overhead production problem lamp lights up and a buzzer sounds. (See Figure 9-10) Now it is the manager's job to solve the problem quickly and take thorough preventive action, so the line will not be stopped again by the same problem.

Problem prevention must be part of an overall management strategy. The TSS slogans for prevention are "rush to the scene" and "on-the-spot confirmation."

- Supervisors with line responsibility must always be present so they can "rush to the scene" as soon as the line is stopped.
- Departments that may be involved in or affected by a problem must be prepared to send someone to the site for "on-the-spot confirmation" as soon as a stop is signalled.

Every line stop is documented on the TSS Confirmation Form. (Figure 9-11) The defect is described, and its cause(s) are confirmed by personnel from the departments involved or affected by the problem. The document must also include an explanation of the temporary or emergency measures taken, the permanent measures that are planned, and who is responsible. This information is circulated to section and division chiefs, as well as the factory QA office. The QA office follows up to determine whether the measures taken are effective and then returns the report to the responsible section. Following this procedure ensures that actions taken will be fact-based. And documenting the process throughout makes it possible to eliminate recurrent problems systematically.

The Results of TSS

Once the TSS system was fully implemented, it began to fulfill its functions smoothly. Problems were discovered more quickly, countermeasures from responsible departments were implemented promptly, and lines were stabilized earlier. With the introduction of TSS, line stoppage time became one of the daily management indicators. Over time, the average line stoppage time per day decreased. (Figure 9-12)

Line Section		TSS Confirmation Form		
Date:	Time Stopped:	Location:	Approved by:	Confirmed by:

Defect Description:

Section Data:

Section_____   Chief_____   Asst. Chief_____

Temporary Measures:	Cause of Defect:
(Immediate)	
1. Reconfirmation	
Who:	
Results:	
2. Action in this Section	
Who:	
Results:	
3. Action in other Sections	
Who:	
Reasons:	

Permanent Measures: Date implemented: Person in charge:

Follow-up on Permanent Measures:

1. Person in charge:
2. Expected date of completion:

Routing: Person making confirmation → Sect. Chief → Division Chief →
QA Office → Sect. Chief

Figure 9-11 NEW TSS Confirmation Form

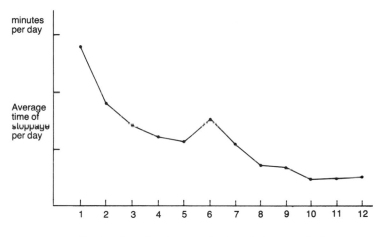

Number of months since commencement of mass production

Figure 9-12 Line Stoppage Time for Product A

PRELIMINARY CONSULTATION SYSTEM FOR SUPPLIERS

At Canon, 70 to 80 percent of all parts come from outside suppliers. So, it is no exaggeration to say that the quality of outside-supplier parts determines the quality of Canon products. For this reason, the selection of cooperating factories (suppliers who cooperate with Canon's quality effort) is a crucial matter. Furthermore, when placing orders with these suppliers, preliminary consultation regarding the quality specifications is also important.

Canon's quality assurance personnel work closely with cooperating factories to make available to them the improvement methods and quality management techniques developed under CPS. They coordinate their efforts with the Outside Order Department, which is also responsible for transferring production assurance techniques to suppliers.

Overall Evaluation of Suppliers

The company's quality requirements are compromised if an order places demands on a cooperating factory that exceed its capacity.

On-site Investigation/Overall Quality Evaluation for Cooperating Factories

(Effective for 1 year)

Date of Investigation:

Code No. | Company

Corporation

Representative	Main Factory		(2nd Factory)				Area Sq. Ft.:	No. Buildings	Employees			Date established	Date began doing business
	Location		Location						Male: Female: Total:				

	Other camera manufacturers	Owned capital	Total capital	Operating profit	Annual sales	Sales to Canon monthly	Canon's degree of dependence	Tamagawa	Purchasing	Shimomaruko	Kosugi	Toride	Fukushima	Utsunomiya	Affiliated companies
Largest customer															

Production System

Equipment	Major machinery
	Major equipment
	Special features of machines, equipment

A	Items to be Investigated	Delivery Record: Year:		From (month):		To (month):			Evaluation Points (a)					b	Rating a × b
		Month	Month	Month	Month	Month	Month	Average	5	4	3	2	1	Weight	Score
Quality of Delivery Time Evaluation Points	Lot fulfillment rate														
	Lot delay rate														

Items to be checked (Is the company increasing profit satisfactorily and operating in a secure manner?)

	Changes over last 5 years	5	4	3	2	1

B	Operating Balance	Profit Rate	Items to be checked	
	(Profitability)	Total capital profit rate	Annual operating profit/ total capital × 100	(Combined result of all operational activities— the higher the better.)

C — Operation

Item	Factory State-of-Health Checklist	5	4	3	2	1	Weight	Score
Operation Strategy	Long-range operating strategy?						4	
Operations Management Indicators	Challenging goals set based on fundamental operational aims?						2	

D — Equipment

Item	Factory State-of-Health Checklist	5	4	3	2	1	Weight	Score
Equipment for labor reduction and rationalization	Annual research and equipment budgets not exceeded?						4	
Capacity to respond to increase/decrease in production	Equipment and reserve capacity managed effectively?						2	

E — Technology

Item	Factory State-of-Health Checklist	5	4	3	2	1	Weight	Score
Unique Technology	Possesses technology that is not duplicated by other companies?						4	

F — Management

Item	Factory State-of-Health Checklist	5	4	3	2	1	Weight	Score
Materials Management	Orderly storage under good conditions and well labeled?						1	
Lead time	Daily-quantity delivery is possible?						4	

G — Management

Item	Factory State-of-Health Checklist	5	4	3	2	1		
Education for human resources development	Implementation of QC education, etc.?						2	
Self development	Self development encouraged through meetings and training?						2	

H — Quality Assurance

Item	Factory State-of-Health Checklist	5	4	3	2	1		
Preliminary consultation regarding preparation for startup	No entries in the Preliminary Agreement are omitted?						4	
Quality Assurance Personnel	QA person registered at Canon as technical specialist in quality control?						2	
Measuring Instrument Control and maintenance	Control regs. established, quick reference chart posted for adjustments?						3	
Adjustment: Examination – Standardization	Sched. checks/adjustments carried out according to standards?						2	
Work Standards	All parts for which orders received standardized with no entries omitted?						4	

Factory Involved in QA Promotion, Beginning 19___

A	B	C	D	E	F	G	H
200	150	95	120	40	95	60	240

PX	MO		1000
		=	10

		SF
Capacity to accept one lump orders		
Has same equipment capacity of this company		

Rank Based on Total Score

	Evaluation	Evaluation	Acknowledgement
A	35 or more		
B	70-84		
C	60-69		
D	40-59		
E	30 or less		

Total score

Figure 9-13 Sample On-Site Quality Evaluation

For this reason, on-the-spot inspection and overall evaluation of suppliers are conducted at fixed intervals. These cover such areas as quality, delivery time, management, equipment, controls, and quality assurance (Figure 9-13). The factories are ranked so that appropriate suppliers can be selected and orders can be made without exceeding a cooperating factory's current capacity.

Preliminary Consultation

Before any order is placed, a preliminary consultation is held between the supplier selected as a cooperating factory and personnel from Canon's ordering, production technology, quality assurance, and testing departments. Prior to the consultation, the supplier is given a preliminary consultation sheet that provides detailed information concerning the parts to be ordered, e.g., parts numbers, name, process, cost, quantity, and shipment size. The supplier then prepares a study of its equipment, manufacturing procedures and methods, manufacturing standards, etc. At the preliminary consultation, this information serves as a starting point for resolving any problems or points that are unclear.

The following items related to quality assurance are thoroughly discussed and agreed upon:

- functions and demanded quality of the product
- use and adjustment of measuring instruments and jigs
- methods for gathering quality data
- acceptance testing standards
- handling and packaging
- persons to contact in case of problems

Quality Assurance Guidance and Assistance

The cooperating factory draws up work standards, carries out the work in accordance with those standards, and confirms quality by means of in-process testing. Canon personnel provide guidance and assistance in the use of statistical methods, controls, and improvement techniques during regularly scheduled discussions or visits to the factory.

Over time, the cooperating factories that develop a solid quality assurance system and score high consistently in acceptance tests are given special recognition as "control test subject factories."

Keys to Fundamental Improvement

Quality assurance means absolute dedication to defect detection and prevention — at the source of defects.

To achieve this requires planning and standardization in all aspects of production, from production technology and work methods to inspection procedures and measurement control. Even more important is every employee's commitment to follow established standards and procedures and to take immediate action whenever problems occur. Comprehensive inspection helps promote adherence. Improvement activities integrated into daily work, such as TSS and the process capability drive, help establish patterns for successful problem solving. Finally, to thoroughly prevent quality defects at their source, the company's quality assurance effort must be extended to suppliers as well.

Reliability in quality must be balanced with reasonable costs. Cost assurance — through cost-control and cost-reduction activities in design and production departments — is the third and final focus of waste elimination activity in the Canon Production System.

10

Cost Assurance Through Value Engineering Activities

CONTINUOUS COST CONTROL

Cost-reduction and cost-control efforts are indispensable in manufacturing. Even before the inauguration of CPS, Canon departments cooperated in ongoing activities related to the continuous cost control cycle. (Figure 10-1)

Cost reductions can be made through either improvements in products or improvements in production methods. The second approach includes automation, labor saving, work improvement, increased equipment efficiency or work rate, reduced expenses,increased yield, and other activities carried out during production, both inside and outside the company. These are all important factors in cost reduction. Since they have been discussed earlier, however, this chapter will focus on product improvement, through value engineering activities in staff departments.

Factory Profit and Product Manufacturing Cost Control

Canon has adopted a standard manufacturing cost system. At the end of the year, every factory reviews improvement

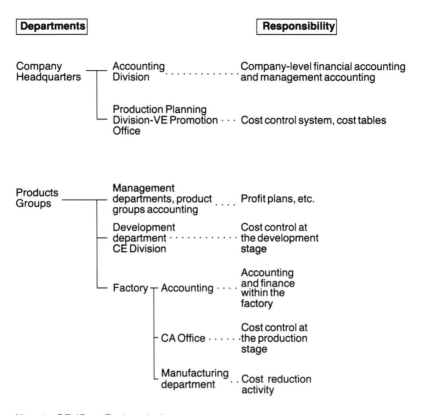

Note 1: CE (Cost Engineering)

Note 2: In some products groups the CE Division and CA office do not exist as separate organizations. In such products groups, personnel responsible for costs in development departments belong to planning sections and personnel responsible for costs in factory departments belong to product technology sections.

Figure 10-1 Departments Concerned with Costs

targets for the coming year and calculates a trial standard manufacturing cost for each product. These trial calculations are then discussed by factory and management departments within the product groups.

After adjustments are made, the standard manufacturing cost for the coming year is established. These standard cost figures and the improvement plans for the same period serve as the basis for the *revaluation amount* (a percentage reduction)

for the standard manufacturing cost. This is an item for adjustment between the factory and management departments in the
product groups. Factory accounting and the management departments compare their plans and calculate how much the standard costs are likely to be reduced. (The revaluation amount is,
in effect, an advance on projected cost-reduction activity. The
lowered standard manufacturing cost *after adjustment* is used
as the account-settlement value for the coming year.)

If improvement efforts during the year produce targeted
results or better and cost-reduction goals are achieved, there is
a profit for the factory. If, on the other hand, improvements cannot be carried out as planned, the factory comes out in the red.
Whether the factory earns a profit or ends up in the red is an
easy index to understand. It helps focus everyone's efforts on
cost-reduction activities.

Control of Product Manufacturing Cost

Factory profit depends on product manufacturing cost
control. To control and reduce costs, cost-related improvement
plans for each product are set up at the end of every year for implementation in the next. Originally, this planning was carried
out solely within the factories. With the advent of CPS and CDS
(Canon Development System), however, the product groups
became involved. Now product designers take part regularly in
brainstorming "retreats" held to generate cost-reduction ideas.

Cost Tables

Cost tables are a simple but important tool for cost control
at Canon. Their function is to standardize prices for supplier
parts and hold them within appropriate and rational levels. The
tables are maintained and adjusted to reflect advances in production technology and manufacturing methods. They are particularly detailed for consigned goods like press and lathe work
and high-frequency orders such as plastics. By analyzing discrepancies between the conditions established in the cost tables
and suppliers' actual situations, Canon can help introduce

improvements at cooperating factories. This is just one of many avenues to improvement at cooperating factories. The factories are visited regularly, and twice a year their top managers attend study meetings to review Canon's productivity and quality improvement activities.

The success of cost-reduction activities depends in part on people's ability to make cost estimates quickly and easily. For this reason, cost tables at Canon are currently being computerized. Once calculation time is reduced, the tables can be used by a much wider group of people — by product designers, factory technologists, and others.

DEVELOPING RELIABLE COST ASSURANCE UNDER CPS

The CPS Committee to Promote Waste Elimination in Planning was given the task of developing a cost assurance system. Its first reaction was probably, "Sure, but what more is there for us to do?" A form of cost control was already being practiced and it was reasonably successful. Fairly high cost-reduction targets had been met every year under difficult conditions. Out of the initial marathon planning sessions, however, a new definition of "assurance" was developed:

- To assure cost means to construct a system in which cost control and cost-reduction activity *produce good results without fail.*
- This does not mean that as long as the results are good the method does not matter. It means putting together a process that *cannot fail to give good results.*

When cost control was examined from this perspective, it became apparent that action was often not taken until after poor results occurred, or only when they were likely to. This tendency to adopt an *ad hoc* or defensive approach to cost control produced disorganization and little, if any, long-term planning. Not surprisingly, it was easy in such an environment to pursue

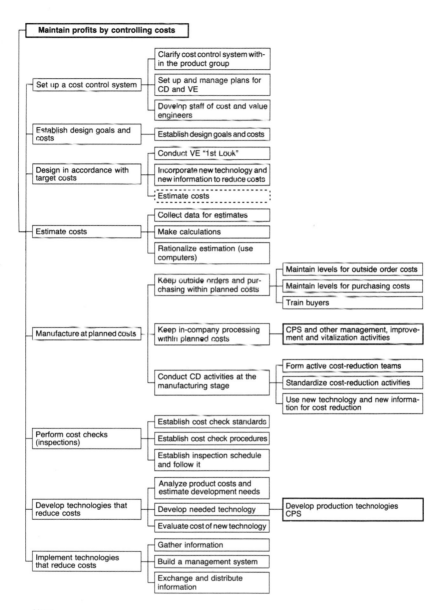

Note:
CD = "cost down," cost-reduction activity
VE 1st Look = Cost reduction before startup in manufacturing (design stage)

Figure 10-2 Organization of Functions to Promote Cost Assurance

ideas based more on wishful thinking than on careful analysis and planning. Many spur-of-the-moment measures proved to be superficial and ineffective over time.

Focusing on problems like these, the group drew up a list of new policies and objectives. First, it made a commitment to start from scratch in preparing a new model — to challenge every established practice in search of better methods. Then it outlined the desired features of the model:

- Establish more organized annual planning and follow-up.
- Strengthen the position of individuals with sole responsibility for cost assurance.
- Reintroduce the functional study method of value engineering.

To fulfill the first of these objectives, a functional distribution chart was drawn up identifying the diverse roles and activities involved in an effective cost assurance system. (Figure 10-2) This closer look at cost-control operations turned out to

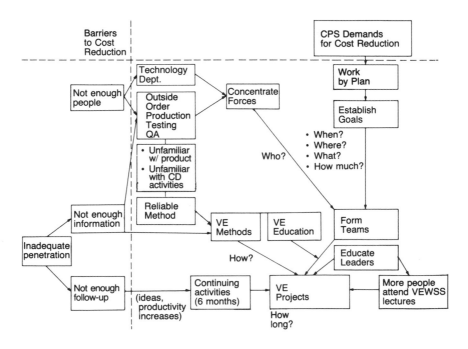

Figure 10-3 The Need for VE in Production Departments

be essential. Relatively few people were assigned to work full-time on costs, and they depended on others to see their cost-reduction and control efforts pay off. But for this very reason, many people in the company did not fully appreciate the importance of cost-control activity. More important, they did not appreciate the significance of their own role in making cost assurance successful. (Figure 10-3)

During this period, cost education was also reevaluated. The committee decided that the cost control system should be supported through human resources development and management training activities. Education and training would make possible two important goals: to give everyone a consistent and correct understanding of costs, and to make accurate cost calculations a fundamental requirement for operation. (See Figure 10-4)

INTRODUCING VE AS A TEAM ACTIVITY

Cost reduction is a team activity involving many people; to make it work, there has to be a framework — an established process that everyone can understand.

Although "start from scratch" may seem like an extreme approach, the committee wanted to create such a framework — a reliable procedure for cost control and especially cost-reduction efforts. Their most important objective was to make sure these activities were *built into* development and production rather than letting them continue on an *ad hoc* basis. After a period of study and observation, the committee chose to adopt the value engineering (VE) process as a basic framework for these activities.

The value engineering process was attractive for several reasons. In VE, an interdisciplinary team measures the current value of a product or its components in terms of functions that fulfill user needs or objectives. For example, a simple mechanical pencil may have eight or more discrete parts in the barrel, spring mechanism, and tip. Each serves a particular function in the overall pencil design, as well as basic user needs. The plastic

Cost Education System

System Objectives:
- To make sure all personnel understand costs
- To make accurate cost calculations a fundamental requirement for all operations

Level	Cost Engineering Staff Development Activities	Training	Basic Cost Education

Manager	• Promote VE and CD
	• Prepare:
	  VE/CD implementation plans
	  Management profit plans for products
	  Overall improvement plans
		• CVS Competency	• VE Training – Sections Chiefs

Supervisor	• Promote improvement activity as VE and CD project leaders • Plan VE and CD projects • Draw up cost tables	• VE 2nd Grade • Competency in other techniques – system engineering, factory management and design, etc.	• Cost Calculation Theory (profit calculation) • Advanced Cost Tables (theory and application of cost tables, special cost tables)
General	• Prepare cost tables, estimates • Participate in VE/CD and other improvement activities • Gain experience in other work centers • Conduct work and standard time studies	• VE 3rd Grade • Product/production technology • Cost engineering • Other techniques – IE, OR, QC, accounting, automation, etc.	• VE 3rd Grade (25% production department indirect personnel) • VE Fundamentals (50% production department indirect personnel) • Basic Cost Education • Advanced Cost Tables

Figure 10-4 Cost Education System

cover on the end of the pencil must provide a firm base for depressing the spring mechanism and propelling the lead. It may also cover or contain an eraser. The relative importance of these functions to the user, or how well the current design of the part fulfills these functions in relation to its cost are some of the questions addressed in functional analysis.

The team develops and evaluates alternatives that might eliminate or improve component areas of low value and matches these new alternatives with the best means of accomplishing them.

The team approach to functional studies was attractive because it seemed likely to encourage involvement; the VE process itself encouraged clear and comprehensive presentation of many ideas. And, since the process is guided by customer preferences, its well-established priorities would be easier for everyone to accept and follow. Finally, VE looked like a reliable framework — the process was already standardized and included a set of well-defined and detailed steps.

Promoting VE Activities

As a management technique, VE has a wide range of applications in many fields. To avoid confusion, however, CPS introduced the value engineering process to the Canon factories in a form referred to as *Second Look VE* — product cost-reduction analysis at the manufacturing stage. (See Figure 10-5) Within the development departments (CDS activities) the VE effort took two forms. In *O (zero) Look VE* product designers used the functional analysis approach in the initial stages of product development, during trial runs at the component level. In *First Look VE,* analysis continues at the planning and trial product stage. [Editor's note: Functional analysis at Canon is a simpler process than that performed in American VE. It is carried out in two steps. First, a *scenario* is developed according to the functions required by or guaranteed to the user. These functions then are classified and organized into a functional (FAST) diagram (called a demanded function diagram). Second, the prod-

0 Look VE Canon System	1st Look VE (Design Stage)	2nd Look VE Canon System
1. Determine the objective–a product strategy based on medium-range plans	1. Determine object–difference between target cost and present conditions	1. Gather Information
2. Estimate costs • Analyze market trends • Analyze similar Canon products • Analyze competing products		
3. Establish target cost		
4. Allocate costs by function (analysis already complete)		
5. Establish cost reduction target amounts by function		
6. Draw up basic outline for the plan		
7. Define functions – Extract (identify) functions (by "scenario" method-using cards) Draw up functional diagram	2. Define functions Extract (identify) functions (Using actual articles or plans) Draw up functional diagram	2. Functional analysis (1) Define top-ranking function (2) Clarify restricting conditions (3) Develop top-ranking function (4) Define functions (5) Organize functions
8. Functional evaluation • Predict costs by function • Evaluate functions • Review cost estimates	3. Functional evaluation • Predict costs by function • Evaluate functions	3. Functional evaluation (1) Predict costs by function (2) Evaluate functions (3) Select target areas
9. Draw up alternatives (1) Generate ideas (2) Evaluate ideas (3) Rank and identify most promising ideas (4) Rank by feasibility (5) Design prototypes based on top ranked functions	4. Draw up alternatives (1) Generate ideas (2) Evaluate ideas (3) Rank and identify promising ideas	4. Draw up alternatives (1) Summarize and evaluate ideas (2) Rank ideas by feasibility (3) Make ideas concrete (4) Refine ideas (5) Draw up alternatives
10. Evaluate cost	5. Evaluate costs	5. Evaluate alternatives
11. Construct, test, and refine	6. Construct, test, and refine	6. Construct, test, and refine
12. Implement and follow up	7. Implement and follow up	7. Proposals and reports

Degree of cost freedom (Possibilities for CD)

Actual Costs

CD = "Cost-down", cost reduction activities

Figure 10-5 Basic Steps of VE Activity

uct parts are classified by their function and incorporated into the diagram.]

Using these various approaches, factory and development departments plan and implement VE according to their own needs. They also cooperate with each other, however, when projects overlap.

The Goals of VE Promotion

After CPS was introduced, the committee's first concrete step was to develop a basic VE training program ("VEWSS" — VE Workshop Seminar) and to revitalize cost-reduction activities. Long- and short-term goals were established:

Long-Term Goals

- Give VE a central role in the cost control system.
- Incorporate cost reductions attributable to VE in profit planning and make sure they are achieved.
- Encourage a function-centered approach to cost reduction throughout the company.

Short-Term Goals

- Use VE concepts and processes in CPS efforts to eliminate waste in planning.
- Introduce cost reduction ("Cost Down") as a team activity that must be carried out continuously.
- Introduce the VE team approach at the development and design stages.

VEWSS — VE Workshop Seminar Program

Canon arranged for the loan of instructors from the Institute of Business Administration to teach that school's 48-hour course in value engineering to Canon personnel. The course focused on Second Look VE for existing company products and emphasized the team approach. VE techniques were taught and applied in actual cost-reduction projects selected by the factory. (Photos 10-1 and 10-2 give an idea of the informal atmosphere of the sessions.)

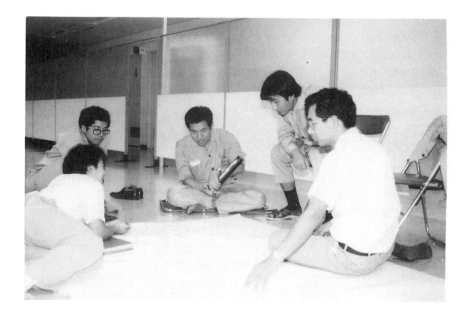

Photo 10-1 Team Learning VE in VEWSS Training (1)

Photo 10-2 Team Learning VE in VEWSS Training (2)

Figure 10-6 History of VE Development at Canon

	'69 '70 '71 '72 '73 '74 '75	'76	'77	'78	'79	'80	'81	'82
Major Developments	**Sept:** Establishment of VA section in the Assets Division	**January:** Lecture meeting (University lecturers) · Lecture meeting · Inauguration of CPS · **July:** Lecture meeting (VE authorities from other companies) · **August:** Development Planning Division joins VE Society		**November:** Production Planning Division joins VE Society	**July:** Canon VE manual drawn up	**January:** VE Promotion Office set up in Production Planning Division · **July:** Establishment of Business Machine CE Center · **November:** 11th VE training sessions for division chiefs		**February:** VE training – I for section chiefs · **November:** VE training – II for section chiefs
VE Training		60 hour VE trainings aimed at products (led by Institute of Business Administration)		Implementation of VE Workshop Seminars (VEWSS)	Aug. → Oct. 3 teams	Apr. → Dec. 10 teams	Apr.→Jun. 5 teams · Aug.→Oct. 6 teams	Feb.→Jul. 10 teams · Nov.→Dec. 7 teams
		Feb.→Apr. · Sep.→Nov. · Mar.→Apr. Total 1.5 months dedicated to VE training				Oct.→Dec. · Aug.→Sep. Total 2.5 months dedicated to VE training		
Individual activities		Implemented among people responsible			Expanded implementation and applications		Integration into ordinary duties	
VEWSS Trainees	50 people / 100 people / 150 people		76: — · 77: 32	78: 41	79: 41	80: 71	81: 133 · 184	82: 278

VEWSS Trainees scale: 50 people, 100 people, 150 people
Year axis: '76 '77 '78 '79 '80 '81 '82

Within three years, 278 people had completed the course in teams of four or five and were awarded diplomas as Value Engineers, Third Grade. Sixty-hour training sessions were held in the product groups over the same period, as well as training sessions for division and section chiefs. (Figure 10-6) The full range of VE education programs is outlined in Figure 10-7. Value engineering activities can be successful as long as at least half the team members understand the process thoroughly. Canon's present objective is to broaden the scope of VE education so that a larger percentage of its personnel will be familar with the VE process.

Course	Audience	Schedule	Lecturers	Content
VE training (division chiefs)	Factory managers	1 day Once a year	Outside consultants Persons promoting VE in other companies	Special lecture on VE and costs
VE training (section chiefs)	Factory/ Development Section shifts	3 days Twice a year	Outside consultants (Institute of Business Administration)	Seminars designed to teach hardware and software aspects of VE. • When and how to use VE for best results • How to involve staff in VE projects
VEWSS	General Goal = train 50% + of relevant staff	2.5 months including 10 days of lecture, 2-3 times a year	Outside consultants (Institute of Business Administration) Company staff	Project topics selected in the factory. Workshop/seminar format (VE 3rd grade awarded)
Advanced VE	People promoting VE in each factory	4 days Once a year	Same as above	Personnel with VE 3rd grade learn essential strategies for awarding VE activities. Goal is to train them as VE project leaders. (VE 2nd grade awarded)
Fundamental VE	General	4 days	Company staff members (VE 2nd grade)	Simple lectures over VE fundamentals and basic background for staff in manufacturing
VE Projects		As needed		Project groups for on-the-job training. These groups also conduct studies aimed at introducing new methods in the company

Figure 10-7 VE Training Programs

VEWSS is the basic CPS training in cost reduction for factory production personnel. Since its subject is cost reduction at the manufacturing stage, medium rather than large-scale improvement is the goal. Participants learn the difference between top-, middle-, and lower-ranking ideas in cost reduction and are encouraged to focus on those that can be implemented

	Top-Ranked Ideas	Middle-Ranked Ideas	Lower-Ranked Ideas
Concept	• Design changes • Method changes • Construction changes greatly	• Construction changes slightly • Assembly chart changes	• Parts diagram changes • Processing method changes
Examples of Ideas	• Method for detecting film movement • Method for coordinating lens and IRED focus positions	• Method for taking up thrust backlash in shaft • Construction of a switch	• Shape of a leaf spring • Shape of a knob
	• Adoption of new materials • Changes in materials		
	Automation, rationalization, and labor-saving in processing method		
Generation Method • Types of Ideas	• Early adoption of advances in technology • Follow-up and completion of research on concept from idea bank	• Functional studies produce additional functions • Ideas for the realization of functions ▼ Aims of VE	• Application of a checklist • Drawing on past experience
Possibilities for Change	Cannot be changed once decision made at start of development	Hard to change after production begins	Can be implemented any time
Handling of VE Results		VE Idea Bank Cost Reduction	

In "0 Look VE" and "1st Look VE," idea generation begins with top-ranking ideas. The entire design process is maintained by repeating the cycle of "construction of basic concept ⟶ detailed design."

Figure 10-8 Second Look VE—Aims of Idea Generation

during production. (See Figure 10-8) This narrower focus helps keep the process simple, manageable, and directed toward application. However, any ideas generated during the training for design stage implementation or further development are forwarded to the idea bank or to the appropriate product development division.

MANAGING VE AND COST-REDUCTION ACTIVITIES

Education and training in VE serve no purpose unless they lead to actual applications and targeted results. Making cost reduction happen in a large company requires "push" as well as planning, especially when operations and conditions differ significantly between factories, production departments, and production-related companies. Even if a uniform approach could be established, time and personnel constraints would prevent its application in many cases. Two key issues were finding the proper *pace* for planning and implementation and developing a management strategy that respected individual differences in target, methods, and depth of application.

Annual Plans for Waste Elimination in Planning

To establish an appropriate pace for planning, unified planning tables were drawn up for CPS waste elimination in planning. (See Figure 10-9) These tables make it possible to see at a glance the continuing cost-reduction activities in every department where CPS had been implemented. In each department, as a rule, idea generation is brought to the concrete stage and completed by the end of the year, then the ideas are put into practice during the following year.

In cost-reduction, the idea-generation stage is only the first step. Ideas must be studied and evaluated for technical feasibility, parts for trial models must be made, and further testing must be done. No proposal can be implemented until there is absolute confirmation that it will work. And, it is essential to have reliable follow-up of this lengthy process.

Subject product	Name of team	Members	Time	Strategy	Target amount
A12 1234	• 2nd Look A12-1234 All units, all parts	(Leaders) 4 People	1982	• Implement ideas generated in 1981 • Evaluate improvements from 1981 • CD meetings for 1982 ideas	Cost reduction amount: $-\$.11/$ unit Total units produced/year: 900 K units Cost reduction amount for year: \$100,000
A12 1235	• 2nd Look A12 1235 All units, all parts	(Leaders) 5 People	1982	• Implement ideas generated in 1981 • CD meetings for 1981 ideas	Cost reduction amount: $-\$1.83/$ unit Total units produced/year: 300K units Cost reduction amount for year: \$550,000
B34 5678	• 1st Look B34 5678 All units, all parts		1982	• 1st Look CD (development stage) completed	Cost reduction amount: $-\$1.00/$ unit Total units produced/year: 100K units Cost reduction amount for year: \$100,000
B34 5679	• 1st Look B34 5679 All units, all parts		1982	• 1st Look CD (development stage) completed	Cost reduction amount: $-\$1.50/$ unit Total units produced/year: 200K units Cost reduction amount for year: \$300,000
B34 5680	• 2nd Look B34 5680 All units, all parts	(Leaders) 4 People	1982	• CD meetings for 1982 ideas • Test ideas generated in VEWSS	Cost reduction amount: $-\$.40/$ unit Total units produced/year 200K units Cost reduction amount for year: \$80,000
(x)	• 1st Look All units, all parts	(Leaders) 7 People	1982	• 1st Look CD at the development stage	Not yet established $= (+\alpha)$

T O T A L:		\$1.13 million
	Sales:	\$17.1 million
	Waste elimination amount/sales:	$2.4 + \alpha$ %

Note: In Section X, CD activities are continuous rather than taking the form of mini-projects. Because of the time frame of implementation, the formula "cost reduction amount for year" does not apply.

Figure 10-9 Planning Chart for CPS Elimination of Waste in Planning

Follow-up methods differ somewhat by factory or division. In many places, however, meetings are held once a month to study actual results, and progress is tracked on large status charts. Seeing their own names listed among the responsible personnel in the annual planning charts has helped to strengthen many individuals' sense of purpose. Figure 10-10 shows annual changes in the elimination of waste in planning since the introduction of CPS.

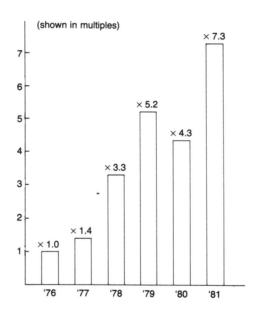

Figure 10-10 Results from Elimination of Waste in Planning

Brainstorming Sessions

Idea generation for cost reduction takes a number of forms, depending on the environment and conditions at a given location. Brainstorming proved to be the most reliable way to encourage involvement and to assure quality participation.

Brainstorming methods include:

- brainstorming within the VE process
- brainstorming on an *ad hoc* basis while looking at plans or products

CPS Technology Transfer

Cost-Reduction Activities for New Products

I want to take this occasion to tell you about the methods and procedures we have been using in our cost reduction activities.

1. Cost reduction before startup of manufacturing (1st Look VE).

   Staff members from the factory side participate at the development stage in task force studies of quality and costs.

   • Estimates, analysis, and evaluation using the design plans · · · · They divide the results of the cost estimates into segments in terms of function, form, and processing and then analyze the differences between estimated costs and target costs, using such techniques as comparison with other products. In this way they are able to identify the problem areas and study possibilities for cost reductions.
   • Marathon cost-cutting sessions · · · · Brainstorming sessions are held, to which specialists from outside the task-force teams are invited. Marathon overnight sessions, lasting three to four days, are repeated five to seven times by the time of manufacturing start-up.
   • The task-force carries out tests and design changes for the sake of cost reduction.
   • The follow-up of cost fluctuations is continued until the cutoff figure to be sent to company headquarters is established.

2. Cost reduction after manufacturing start-up (2nd Look VE)

   Even after start-up of manufacturing the actual figure may be larger than the cutoff figure, so cost reduction must be pushed even further.

   • Formation of cost reduction teams within the factory · · · · In 1st Look VE, the emphasis was on design changes, but in 2nd Look VE the emphasis is on improvements based on analysis and evaluation of current manufacturing costs. Foremen participate as team members, and the united strength of the entire factory is directed at the problem.
   • Monthly meetings to report and study actual results · · · · If ideas are not implemented, they will have no effect. Therefore all sections require progress reports on the plan and on prospects for the future.

3. For the future · · · · Cost reduction is not easy.

   It is essential for everyone concerned to continue making reductions slowly but surely from the development stage onwards. That is the way to get big results in cost reduction!

   • Thorough study of design plans before start-up of manufacturing · · · · Solve as many quality problems at the development stage as possible. Troubles after start-up are a major factor in start-up losses. If quality isn't stabilized, costs don't stabilize.
   • In cost-reducing brainstorming, keep the emphasis on function (VE concepts).
   • Implement cost reductions as quickly as possible.

Figure 10-11 Description of Marathon Brainstorming Sessions from a Company Newsletter

- selecting various themes such as assembly methods, circuitry, or unit construction items and then assigning the themes to groups for brainstorming.

Brainstorming can involve just one person, a group from a section division responsible for product technology, or a mixed group involving departments within the factory or the development division.

Recently, brainstorming retreats have become a popular part of cost-reduction activities. These overnight meetings are usually held in October or November. The participants gather in lodgings inside or outside the company. Then, surrounded by documents and actual products, they engage in discussions and brainstorming sessions late into the night.

This technique has a number of advantages. Since the participants are away from their usual activities, they can concentrate on generating ideas without interference from calls on the PA system or other noise and interruptions. Sessions can be concluded within a relatively short period, and this makes it easier to involve people from other departments. Finally, participants enjoy the work. A sample of that enthusiasm is illustrated in Figure 10-11. This is an excerpt from an article on cost-reduction activities written by an individual with cost responsibilities at one of the factories and published in a company newsletter.

Example of Cost-Reduction Activity

Figure 10-12 illustrates a typical example of cost-reduction activity carried on in the factories. After planning, the factory CA office identified the object of cost reduction, set target figures, and put together the CD ("cost-down") teams. Brainstorming was done by all team members at a retreat session. Follow-up and management of results in this case were handled through monthly meetings to study actual results.

Promoting Participation at the Development Level

All CPS activities are aimed at production, so it is only natural that cost-reduction activities under CPS should focus on "Second Look VE" — improvements that can be implemented after production has started. From a company-wide perspective, however, the further upstream cost-reduction activity can be conducted the more effective it will be.

In some product groups product cost calculations at the development stage are done by factory cost personnel. In these groups, factory cost personnel play a central role in cost-reduction meetings at the development stage. Increasingly, factory personnel are asked to participate in cost-reduction activities at the design stage. To encourage this participation, the results of "First Look VE" are counted as part of overall CPS results whenever factory production personnel participate in the project. The results are counted as part of that factory's CPS profit for one year after startup of production.

Examples of Implemented Improvement Proposals

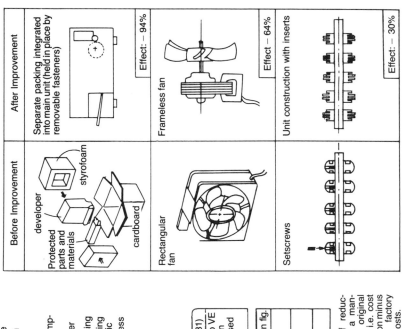

Before Improvement	After Improvement
Protected parts and materials — developer, styrofoam, cardboard	Separate packing integrated into main unit (held in place by removable fasteners) **Effect: – 94%**
Rectangular fan	Frameless fan **Effect – 64%**
Setscrews	Unit construction with inserts **Effect: – 30%**

1. Object Themes

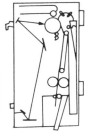

- Product name: NP-A
- Selling price: XXXX
- Side view:

- 1st Look VE is finished (design stage VE)

Special functional features of product:
1. Copy speed is 12 sheets/minute
2. Small-model PPC machine with fixed holder for originals
3. Waiting time 0 seconds
4. Low-energy (low electric consumption) type
5. pressure-fixing on ordinary paper

Special features related to processing
1. Process without cutting or grinding
   Increased use of moulded plastic
   Increased use of plate metal press

2. CD (Cost down) target figures

1st revision (addition) (12/80)
- Addition to insure profit
- Recovery of increase from plan revision

2nd revision (2/81)
Decrease due to VE estimate revision offset by increased productivity

3rd revision (4/81)
Decrease due to VE estimate revision offset by increased productivity

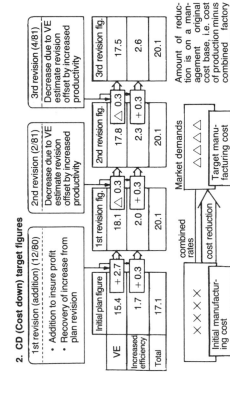

	Initial plan figure		1st revision fig.		2nd revision fig.		3rd revision fig.
VE	15.4	+ 2.7	18.1	△ 0.3	17.8	△ 0.3	17.5
Increased efficiency	1.7	+ 0.3	2.0	+ 0.3	2.3	+ 0.3	2.6
Total	17.1		20.1		20.1		20.1

Market demands △△△△

combined rates

cost reduction

Initial manufacturing cost X X X X

Target manufacturing cost △△△△△

Amount of reduction is on a management original cost base, i.e. cost of production minus combined factory overhead costs.

Unit: ¥1,000

3. Cost Down Schedule

Market needs → Demands for cost reduction → Inauguration of CD teams → First 2-day brainstorm session (14 people) → Second 2-day brainstorm session (12 people)

Members: Factory: 8 people, R/D: 5 people

Review ideas → Evaluation [Ease of repair, Trial manufacture, QA tests] → Plan revision → Implementation

Idea Sources

- Development of functions
- Tables comparing functions to costs
- Idea from VE exhibit room
- Use of VE case studies (idea cards)
- Comparison with machines from other companies
- Review of 1st Look VE
- Ideas from various sections
- Division of responsibility and assignment to groups by unit
- VE proposals for changes in external appearance

5. Follow-up and Management of Results

Changes in CD result amounts and actual manufacturing costs

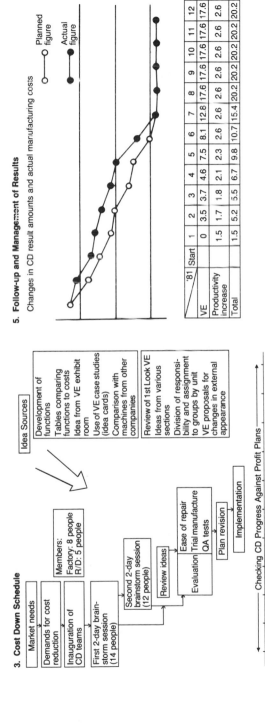

○—○ Planned figure ●—● Actual figure

'81	Start	1	2	3	4	5	6	7	8	9	10	11	12
VE	0	3.5	3.7	4.6	7.5	8.1	12.8	17.6	17.6	17.6	17.6	17.6	17.6
Productivity increase	1.5	1.7	1.8	2.1	2.3	2.6	2.6	2.6	2.6	2.6	2.6	2.6	2.6
Total	1.5	5.2	5.5	6.7	9.8	10.7	15.4	20.2	20.2	20.2	20.2	20.2	20.2

4. Number of Ideas and Rate of Implementation

— Checking CD Progress Against Profit Plans →

'80 11 12 1 2 3 4 5 6 7 8 9

Level of Difficulty

Degree of difficulty	Ideas			Actual Results		
	# Ideas	Feasibility	Estimated cost	# Ideas	Actual cost	Feasibility
A: Trial mfg. div. studies	60	3109 × 0.9	= 2798	44	2107	0.68
B: QA test needs	93	22454 × 0.5	= 11227	50	11837	0.53
C*: QA test needs	52	13785 × 0.2	= 2757	17	3659	0.27
Total	205	39348	= 16782	111	17603	

*Includes C-rank external-appearance changes

6. Development Aimed at Other Types of Machines

A Items adopted for NP-A and also reflected in NP-B
1. Adoption of one-sided AC driver
2. Cooling fan with frameless motor and blade
3. Fewer toner sensors
4. Other   Total × × × yen

B Items not adopted for NP-A but accepted for NP-B
1. Replaced fluorescent with halogen lamp system
2. Replaced main motor worm gear with direct gear   Total ○○○ yen
3. Other   For NP-B: Total △△△ yen

Figure 10-12   Example of VE Activity

KEYS TO FUNDAMENTAL IMPROVEMENT

Eliminating waste reduces costs and adds value to products without capital investment — to increase the company's profit margin.

Adding value through waste elimination is Canon's fundamental strategy for overall improvement, producing cost reductions in every aspect of manufacturing — from product design to delivery. This chapter focused on a single, important aspect of cost assurance — activities to promote product improvement through value engineering. Every improvement activity at Canon, however, is dedicated to reducing costs.

When we look for ways to improve, we may consider new equipment, facilities, and systems — and new or fewer people. The most innovative step of all, however, is to make the most of the resources we have by making improvements that eliminate waste. Giving everyone in the company cost-reduction tools and training lets them think and act with true cost-consciousness. Then, cost-reduction activity becomes part of day-to-day work and management — a continuous improvement strategy — not just one of the measures we take in a crisis.

Index

Other Tools for Continuous Improvement

Productivity Press publishes and distributes materials on continuous improvement in productivity, quality, customer service, and the creative involvement of all employees. Many of our products are direct source materials from Japan that have been translated into English for the first time and are available exclusively from Productivity. Supplemental products and services include newsletters, conferences, seminars, in-house training and consulting, audio-visual training programs, and industrial study missions. Call 1-800-274-9911 for our free book catalog.

TPM Development Program
Implementing Total Productive Maintenance
edited by Seiichi Nakajima

This book outlines a three-year program for systematic TPM development and implementation. It describes in detail the five principal developmental activities of TPM:
1. Systematic elimination of the six big equipment related losses through small group activities
2. Autonomous maintenance (by operators)
3. Scheduled maintenance for the maintenance department
4. Training in operation and maintenance skills
5. Comprehensive equipment management from the design stage
ISBN 0-915299-37-2 / 428 pages / $85.00 / Order code DTPM- BK

Training for TPM: A Manufacturing Success Story
edited by Nachi Fujikoshi

A detailed case study of TPM implementation at a world-class manufacturer of bearings, precision machine tools, dies, industrial equipment, and robots. In 1984, two and a half years after beginning implementation, the company was awarded Japan's prestigious PM Prize for its program. Here's a detailed account of their improvement activities — and an impressive model for yours.
ISBN 0-915299-34-8 / 320 pages / $59.95 / Order code CTPM-BK

Total Productive Maintenance
Maximizing Productivity and Quality (AV)
Japan Management Association

Introduce TPM to your work force in this accessible two-part audio visual program, which explains the rationale and basic principles of TPM to supervisors, group leaders, and workers. It explains five major developmental activities of TPM, includes a section on equipment improvement that focuses on eliminating chronic losses, and describes an analytical approach called PM Analysis to help solve problems that have complex and continuously changing causes. (Approximately 45 minutes long.)
167 Slides / ISBN 0-915299-46-1 / $749.00 / Order code STPM-BK
2 Videos / ISBN 0-915299-49-6 / $749.00 / Order code VTPM-BK

JIT Factory Revolution
A Pictorial Guide to Factory Design of the Future
by Hiroyuki Hirano/JIT Management Library

Here is the first-ever encyclopedic picture book of JIT. With 240 pages of photos, cartoons, and diagrams, this unprecedented behind-the-scenes look at actual production and assembly plants shows you exactly how JIT looks and functions. It shows you how to set up each area of a JIT plant and provides hundreds of useful ideas you can implement. If you've made the crucial decision to run production using JIT and want to show your employees what it's all about, this book is a must. The photographs, from Japanese production and assembly plants, provide vivid depictions of what work is like in a JIT environment. And the text, simple and easy to read, makes all the essentials crystal clear.
ISBN 0-915299-44-5 / 240 pages / $49.95 / Order code JITFAC-BK

JIT Implementation Manual
The Complete Guide to Just-In-Time Manufacturing
by Hiroyuki Hirano

Here is the most comprehensive and detailed manual we have found anywhere for setting up a complete JIT program. Encyclopedic in scope, and written by a top international consultant, it provides the JIT professional with the answer to virtually any JIT problem. It shows multiple options for handling every stage of implementation and is appropriate to all factory settings, whether in job shop, repetitive, or process manufacturing. Covering JIT concepts, techniques, and tools, and including hundreds of illustrations, charts, diagrams, and JIT management forms, this manual is a truly indispensable tool.
ISBN 0-915299-66-6/1000 + pages in 2 volumes/$3500.00/Order code HIRJIT-BK

Kanban and Just-In-Time at Toyota
Management Begins at the Workplace (rev.)
edited by the Japan Management Association, translated by David J. Lu

Based on seminars developed by Taiichi Ohno and others at Toyota for their major suppliers, this book is the best practical introduction to Just-In-Time available. Now in a newly expanded edition, it explains every aspect of a "pull" system in clear and simple terms — the underlying rationale, how to set up the system and get everyone involved, and how to refine it once it's in place. A groundbreaking and essential tool for companies beginning JIT implementation.
ISBN 0-915299-48-8 / 224 pages / $36.50 / Order code KAN-BK

Total Manufacturing Management
by Giorgio Merli

One of Italy's leading consultants discusses the implementation of Just-In-Time and related methods (including QFD and TPM) in Western corporations. The author does not approach JIT from a mechanistic orientation aimed simply at production efficiency. Rather, he discusses JIT from the perspective of industrial strategy and as an overall organizational model. Here's a sophisticated program for organizational reform that shows how JIT can be applied even in types of production that have often been neglected in the West, including custom work.
ISBN 0-915299-58-5 / 224 pages / $39.95 / Order code TMM-BK

A Study of the Toyota Production System
From an Industrial Engineering Viewpoint (rev.)
by Shigeo Shingo

The "green book" that started it all — the first book in English on JIT, now completely revised and re-translated. Here is Dr. Shingo's classic industrial engineering rationale for the priority of process-based over operational improvements for manufacturing. He explains the basic mechanisms of the Toyota production system in a practical and simple way so that you can apply them in your own plant.
ISBN 0-915299-17-8 / 352 pages / Price $39.95 / Order code STREV-BK

Productivity Press, Inc., Dept. BK, P.O. Box 3007, Cambridge, MA 02140 1-800-274-9911

The Sayings of Shigeo Shingo
Key Strategies for Plant Improvement
by Shigeo Shingo, translated by Andrew P. Dillon

A recent issue of Quality Digest claims that Shigeo Shingo "is an unquestioned genius — the Thomas Edison of Japan." The author, a world-renowned expert on manufacturing, "offers new ways to discover the root causes of manufacturing problems. These discoveries can set in motion the chain of cause and effect, leading to greatly increased productivity." By means of hundreds of fascinating real-life examples, Shingo describes many simple ways to identify, analyze, and solve problems in the workplace. This is an accessible, readable, and helpful book for anyone who wants to improve productivity.
ISBN 0-915299-15-1 / 208 pages / $39.95 / Order code SAY-BK

A Revolution in Manufacturing
The SMED System
by Shigeo Shingo, translated by Andrew P. Dillon

SMED (Single-Minute Exchange of Die), or quick changeover techniques, is the single most powerful tool for Just-In-Time production. Written by the industrial engineer who developed SMED for Toyota, the book contains hundreds of illustrations and photographs, as well as twelve chapter-length case studies. Here are the most complete and detailed instructions available anywhere for transforming a manufacturing environment to speed up production (Shingo's average setup time reduction is an astounding 98 percent) and make small-lot inventories feasible.
ISBN 0-915299-03-8 / 383 pages / $70.00 / Order code SMED-BK

Zero Quality Control
Source Inspection and the Poka-yoke System
by Shigeo Shingo, translated by Andrew P. Dillon

A remarkable combination of source inspection (to detect errors before they become defects) and mistake-proofing devices (to weed out defects before they can be passed down the production line) eliminates the need for statistical quality control. Shingo shows how this proven system for reducing defects to zero turns out the highest quality products in the shortest period of time. With over 100 specific examples illustrated. (Audio-visual training program also available.)
ISBN 0-915299-07-0 / 328 pages / $70.00 / Order code ZQC-BK

Productivity Press, Inc., Dept. BK, P.O. Box 3007, Cambridge, MA 02140 1-800-274-9911

Non-Stock Production
The Shingo System for Continuous Improvement
by Shigeo Shingo

Shingo, whose work at Toyota provided the foundation for JIT, teaches how to implement non-stock production in your JIT manufacturing operations. The culmination of his extensive writings on efficient production management and continuous improvement, this book is an essential companion volume to his other landmark books on key elements of JIT, including SMED and Poka-Yoke.
ISBN 0-915299-30-5 / 480 pages / $75.00 / Order code NON-BK

The Improvement Book
Creating the Problem-Free Workplace
by Tomo Sugiyama

A practical guide to setting up a participatory problem-solving system in the workplace. Focusing on ways to eliminate the "Big 3" problems — irrationality, inconsistency, and waste — this book provides clear direction for starting a "problem-free engineering" program. It also gives you a full introduction to basic concepts of industrial housekeeping (known in Japan as 5S), two chapters of examples that can be used in small group training activities, and a workbook for individual use (extra copies are available separately). Written in an informal style, and using many anecdotes and examples, this book provides a proven approach to problem solving for any industrial setting.
ISBN 0-915299-47-X / 236 pages / $49.95 / Order code IB-BK

Variety Reduction Program (VRP)
A Production Strategy for Product Diversification
by Toshio Suzue and Akira Kohdate

Here's the first book in English on a powerful way to increase manufacturing flexibility without increasing costs. How? By reducing the number of parts within each product type and by simplifying and standardizing parts between models. VRP is an integral feature of advanced manufacturing systems. This book, an introduction to and handbook for VRP implementation, features over 100 illustrations and is written for top manufacturing executives, middle managers, and R&D personnel.
ISBN 0-915299-32-1 / 164 pages / $59.95 / Order code VRP-BK

Productivity Press, Inc., Dept. BK, P.O. Box 3007, Cambridge, MA 02140 1-800-274-9911

Quick Changeover Users Group

Quick changeover, developed by Shigeo Shingo at Toyota, is the heart of JIT. Whether you're a novice or veteran of quick changeover techniques, you'll gain more information, more tools for implementation, and more applications of this powerful productivity bookster by joining the Quick Changeover User Group. Benefits of membership include a monthly newsletter, special rates on conferences and seminars, and discounts on selected books. Annual fee is $357. For more information about membership and benefits, call 1-800-888-6485.

TO ORDER: Write, phone, or fax Productivity Press, Dept. BK, P.O. Box 3007, Cambridge, MA 02140, phone 1-800-274-9911, fax 617-868-3524. Send check or charge to your credit card (American Express, Visa, MasterCard accepted).

U.S. ORDERS: Add $4 shipping for first book, $2 each additional for UPS surface delivery. Add $10 for each AV program you order. CT residents add 8% and MA residents 5% sales tax. We offer attractive quantity discounts for bulk purchases of individual titles; call for more information.

FOREIGN ORDERS: Pre-payment in U.S. dollars must accompany your order (checks must be drawn on U.S. banks). For international orders write, phone, or fax for quote and indicate shipping method desired. When quote is returned with payment, your order will be shipped promptly by the method requested.

NOTE: Prices subject to change without notice.

COMPLETE LIST OF TITLES FROM PRODUCTIVITY PRESS

Akao, Youji (ed.). **Quality Function Deployment: Integrating Customer Requirements into Product Design**
ISBN 0-915299-41-0 / 1990 / 320 pages / $75.00 / Order code QFD-BK

Asaka, Tetsuichi and Kazuo Ozeki (eds.). **Handbook of Quality Tools: The Japanese Approach**
ISBN 0-915299-45-3 / 1990 / 320 pages / $59.95 / order code HQT-BK

Belohlav, James A. **Championship Management**
ISBN 0-915299-76-3 / 1990 / 176 pages / $29.95 / Order code CHAMPS-BK

Christopher, William F. **Productivity Measurement Handbook**
ISBN 0-915299-05-4 / 1985 / 680 pages / $137.95 / order code PMH-BK

D'Egidio, Franco. **Global Service Management**
ISBN 0-915299-68-2 / 1990 / 194 pages / $34.95 / Order code GSM-BK

Ford, Henry. **Today and Tomorrow**
ISBN 0-915299-36-4 / 1988 / 286 pages / $24.95 / order code FORD-BK

Fukuda, Ryuji. **CEDAC: A Tool for Continuous Systematic Improvement**
ISBN 0-915299-26-7 / 1990 / 144 pages / $49.95 / order code CEDAC-BK

Fukuda, Ryuji. **Managerial Engineering: Techniques for Improving Quality and Productivity in the Workplace**
ISBN 0-915299-09-7 / 1984 / 206 pages / $39.95 / order code ME-BK

Hatakeyama, Yoshio. **Manager Revolution! A Guide to Survival in Today's Changing Workplace**
ISBN 0-915299-10-0 / 1985 / 208 pages / $24.95 / order code MREV-BK

Hirano, Hiroyuki. **JIT Factory Revolution: A Pictorial Guide to Factory Design of the Future**
ISBN 0-915299-44-5 / 1989 / 227 pages / $49.95 / order code JITFAC-BK

Hirano, Hiroyuki. **JIT Implementation Manual: The Complete Guide to Just-In-Time Manufacturing**
ISBN 0-915299-66-6 / 1990 / 1000 + pages / $3500.00 / order code HIRJIT-BK

Horovitz, Jacques. **Winning Ways: Achieving Zero Defects Service**
ISBN 0-915299-78-X / 1990 / 176 pages / $24.95 / Order code WWAYS-BK

Japan Human Relations Association (ed.). **The Idea Book: Improvement Through TEI (Total Employee Involvement)**
ISBN 0-915299-22-4 / 1988 / 232 pages / $49.95 / order code IDEA-BK

Japan Human Relations Association (ed.). **The Idea Book for Customer Service**
ISBN 0-915299-65-8 / 1990 / 272 pages / $49.95 / Order code QSIDEA-BK

Japan Management Association (ed.). **Kanban and Just-In-Time at Toyota: Management Begins at the Workplace** (Revised Ed.), Translated by David J. Lu
ISBN 0-915299-48-8 / 1989 / 224 pages / $36.50 / order code KAN-BK

Japan Management Association and Constance E. Dyer. **The Canon Production System: Creative Involvement of the Total Workforce**
ISBN 0-915299-06-2 / 1987 / 251 pages / $36.95 / order code CAN-BK

Jones, Karen (ed.). **The Best of TEI: Current Perspectives on Total Employee Involvement**
ISBN 0-915299-63-1 / 1989 / 502 pages / $175.00 / order code TEI-BK

Karatsu, Hajime. **Tough Words For American Industry**
ISBN 0-915299-25-9 / 1988 / 178 pages / $24.95 / order code TOUGH-BK

Karatsu, Hajime. **TQC Wisdom of Japan: Managing for Total Quality Control**,
Translated by David J. Lu
ISBN 0-915299-18-6 / 1988 / 136 pages / $34.95 / order code WISD-BK

Kobayashi, Iwao. **20 Keys to Workplace Improvement**
ISBN 0-915299-61-5 / 1990 / 224 pages / $34.95 / Order code 20KEYS-BK

Lu, David J. **Inside Corporate Japan: The Art of Fumble-Free Management**
ISBN 0-915299-16-X / 1987 / 278 pages / $24.95 / order code ICJ-BK

Merli, Giorgio. **Total Manufacturing Management**
ISBN 0-915299-58-5 / 1990 / pages / $39.95 / Order code TMM-BK

Mizuno, Shigeru (ed.). **Management for Quality Improvement: The 7 New
QC Tools**
ISBN 0-915299-29-1 / 1988 / 324 pages / $59.95 / order code 7QC-BK

Monden, Yashuhiro and Michiharu Sakurai (eds.). **Japanese Management
Accounting: A World Class Approach to Profit Management**
ISBN 0-915299-50-X / 1989 / 584 pages / $59.95 / order code JMACT-BK

Nachi Fujikoshi (ed.). **Training for TPM: A Manufacturing Success Story**
ISBN 0-915299-34-8 / 1990 / 320 pages / $59.95 / Order code CTPM-BK

Nakajima, Seiichi. **Introduction to TPM: Total Productive Maintenance**
ISBN 0-915299-23-2 / 1988 / 149 pages / $39.95 / order code ITPM-BK

Nakajima, Seiichi. **TPM Development Program: Implementing Total
Productive Maintenance**
ISBN 0-915299-37-2 / 1989 / 428 pages / $85.00 / order code DTPM-BK

Nikkan Kogyo Shimbun, Ltd./Factory Magazine (ed.). **Poka-yoke: Improving
Product Quality by Preventing Defects**
ISBN 0-915299-31-3 / 1989 / 288 pages / $59.95 / order code IPOKA-BK

Ohno, Taiichi. **Toyota Production System: Beyond Large-Scale Production**
ISBN 0-915299-14-3 / 1988 / 162 pages / $39.95 / order code OTPS-BK

Ohno, Taiichi. **Workplace Management**
ISBN 0-915299-19-4 / 1988 / 165 pages / $34.95 / order code WPM-BK

Ohno, Taiichi and Setsuo Mito. **Just-In-Time for Today and Tomorrow**
ISBN 0-915299-20-8 / 1988 / 165 pages / $34.95 / order code OMJIT-BK

Psarouthakis, John. **Better Makes Us Best**
ISBN 0-915299-56-9 / 1989 / 112 pages / $16.95 / order code BMUB-BK

Robson, Ross (ed.). **The Quality and Productivity Equation: American
Corporate Strategies for the 1990s**
ISBN 0-915299-71-2 / 1990 / 558 pages / $29.95 / order code QPE-BK

Shetty, Y.K and Vernon M. Buehler (eds.). **Competing Through Productivity
and Quality**
ISBN 0-915299-43-7 / 1989 / 576 pages / $39.95 / order code COMP-BK

Shingo, Shigeo. **Non-Stock Production: The Shingo System for Continuous
Improvement**
ISBN 0-915299-30-5 / 1988 / 480 pages / $75.00 / order code NON-BK

Productivity Press, Inc., Dept. BK, P.O. Box 3007, Cambridge, MA 02140 1-800-274-9911

Shingo, Shigeo. **A Revolution In Manufacturing: The SMED System**,
Translated by Andrew P. Dillon
ISBN 0-915299-03-8 / 1985 / 383 pages / $70.00 / order code SMED-BK

Shingo, Shigeo. **The Sayings of Shigeo Shingo: Key Strategies for Plant
Improvement**, Translated by Andrew P. Dillon
ISBN 0-915299-15-1 / 1987 / 208 pages / $39.95 / order code SAY-BK

Shingo, Shigeo. **A Study of the Toyota Production System from an Industrial
Engineering Viewpoint** (Revised Ed.),
ISBN 0-915299-17-8 / 1989 / 293 pages / $39.95 / order code STREV-BK

Shingo, Shigeo. **Zero Quality Control: Source Inspection and the Poka-yoke
System**, Translated by Andrew P. Dillon
ISBN 0-915299-07-0 / 1986 / 328 pages / $70.00 / order code ZQC-BK

Shinohara, Isao (ed.). **New Production System: JIT Crossing Industry
Boundaries**
ISBN 0-915299-21-6 / 1988 / 224 pages / $34.95 / order code NPS-BK

Sugiyama, Tomo. **The Improvement Book: Creating the Problem-free
Workplace**
ISBN 0-915299-47-X / 1989 / 236 pages / $49.95 / order code IB-BK

Suzue, Toshio and Akira Kohdate. **Variety Reduction Program (VRP): A
Production Strategy for Product Diversification**
ISBN 0-915299-32-1 / 1990 / 164 pages / $59.95 / order code VRP-BK

Tateisi, Kazuma. **The Eternal Venture Spirit: An Executive's Practical
Philosophy**
ISBN 0-915299-55-0 / 1989 / 208 pages / $19.95 / order code EVS-BK

AUDIO-VISUAL PROGRAMS

Japan Management Association. **Total Productive Maintenance: Maximizing
Productivity and Quality**
ISBN 0-915299-46-1 / 167 slides / 1989 / $749.00 / order code STPM
ISBN 0-915299-49-6 / 2 videos / 1989 / $749.00 / order code VTPM

Shingo, Shigeo. **The SMED System**, Translated by Andrew P. Dillon
ISBN 0-915299-11-9 / 181 slides / 1986 / $749.00 / order code S5
ISBN 0-915299-27-5 / 2 videos / 1987 / $749.00 / order code V5

Shingo, Shigeo. **The Poka-yoke System**, Translated by Andrew P. Dillon
ISBN 0-915299-13-5 / 235 slides / 1987 / $749.00 / order code S6
ISBN 0-915299-28-3 / 2 videos / 1987 / $749.00 / order code V6

Productivity Press, Inc., Dept. BK, P.O. Box 3007, Cambridge, MA 02140 1-800-274-9911